MULTI-UNIT PROBABILISTIC SAFETY ASSESSMENT

The following States are Members of the International Atomic Energy Agency:

AFGHANISTAN	GEORGIA	PAKISTAN
ALBANIA	GERMANY	PALAU
ALGERIA	GHANA	PANAMA
ANGOLA	GREECE	PAPUA NEW GUINEA
ANTIGUA AND BARBUDA	GRENADA	PARAGUAY
ARGENTINA	GUATEMALA	PERU
ARMENIA	GUYANA	PHILIPPINES
AUSTRALIA	HAITI	POLAND
AUSTRIA	HOLY SEE	PORTUGAL
AZERBAIJAN	HONDURAS	QATAR
BAHAMAS	HUNGARY	REPUBLIC OF MOLDOVA
BAHRAIN	ICELAND	ROMANIA
BANGLADESH	INDIA	RUSSIAN FEDERATION
BARBADOS	INDONESIA	RWANDA
BELARUS	IRAN, ISLAMIC REPUBLIC OF	SAINT KITTS AND NEVIS
BELGIUM	IRAQ	SAINT LUCIA
BELIZE	IRELAND	SAINT VINCENT AND
BENIN	ISRAEL	THE GRENADINES
BOLIVIA, PLURINATIONAL	ITALY	SAMOA
STATE OF	JAMAICA	SAN MARINO
BOSNIA AND HERZEGOVINA	JAPAN	SAUDI ARABIA
BOTSWANA	JORDAN	SENEGAL
BRAZIL	KAZAKHSTAN	SERBIA
BRUNEI DARUSSALAM	KENYA	SEYCHELLES
BULGARIA	KOREA, REPUBLIC OF	SIERRA LEONE
BURKINA FASO	KUWAIT	SINGAPORE
BURUNDI	KYRGYZSTAN	SLOVAKIA
CAMBODIA	LAO PEOPLE'S DEMOCRATIC	SLOVENIA
CAMEROON	REPUBLIC	SOUTH AFRICA
CANADA	LATVIA	SPAIN
CENTRAL AFRICAN	LEBANON	SRI LANKA
REPUBLIC	LESOTHO	SUDAN
CHAD	LIBERIA	SWEDEN
CHILE	LIBYA	SWITZERLAND
CHINA	LIECHTENSTEIN	SYRIAN ARAB REPUBLIC
COLOMBIA	LITHUANIA	TAJIKISTAN
COMOROS	LUXEMBOURG	THAILAND
CONGO	MADAGASCAR	TOGO
COSTA RICA	MALAWI	TONGA
CÔTE D'IVOIRE	MALAYSIA	TRINIDAD AND TOBAGO
CROATIA	MALI	TUNISIA
CUBA	MALTA	TÜRKİYE
CYPRUS	MARSHALL ISLANDS	TURKMENISTAN
CZECH REPUBLIC	MAURITANIA	UGANDA
DEMOCRATIC REPUBLIC	MAURITIUS	UKRAINE
OF THE CONGO	MEXICO	UNITED ARAB EMIRATES
DENMARK	MONACO	UNITED KINGDOM OF
DJIBOUTI	MONGOLIA	GREAT BRITAIN AND
DOMINICA	MONTENEGRO	NORTHERN IRELAND
DOMINICAN REPUBLIC	MOROCCO	UNITED REPUBLIC
ECUADOR	MOZAMBIQUE	OF TANZANIA
EGYPT	MYANMAR	UNITED STATES OF AMERICA
EL SALVADOR	NAMIBIA	URUGUAY
ERITREA	NEPAL	UZBEKISTAN
ESTONIA	NETHERLANDS	VANUATU
ESWATINI	NEW ZEALAND	VENEZUELA, BOLIVARIAN
ETHIOPIA	NICARAGUA	REPUBLIC OF
FIJI	NIGER	VIET NAM
FINLAND	NIGERIA	YEMEN
FRANCE	NORTH MACEDONIA	ZAMBIA
GABON	NORWAY	ZIMBABWE
GAMBIA	OMAN	

The Agency's Statute was approved on 23 October 1956 by the Conference on the Statute of the IAEA held at United Nations Headquarters, New York; it entered into force on 29 July 1957. The Headquarters of the Agency are situated in Vienna. Its principal objective is "to accelerate and enlarge the contribution of atomic energy to peace, health and prosperity throughout the world".

SAFETY REPORTS SERIES No. 110

MULTI-UNIT PROBABILISTIC SAFETY ASSESSMENT

INTERNATIONAL ATOMIC ENERGY AGENCY
VIENNA, 2023

COPYRIGHT NOTICE

© IAEA, 2023

Printed by the IAEA in Austria
September 2023
STI/PUB/1974

IAEA Library Cataloguing in Publication Data

Names: International Atomic Energy Agency.
Title: Multi-unit probabilistic safety assessment / International Atomic Energy Agency.
Description: Vienna : International Atomic Energy Agency, 2023. | Series: IAEA safety reports series, ISSN 1020–6450 ; no. 110 | Includes bibliographical references.
Identifiers: IAEAL 22-01585 | ISBN 978–92–0–119222–6 (paperback : alk. paper) | ISBN 978–92–0–119322–3 (pdf) | ISBN 978–92–0–119422–0 (epub)
Subjects: LCSH: Nuclear power plants — Safety measures. | Nuclear power plants — Risk assessment. | Industrial safety.
Classification: UDC 621.039.58 | STI/PUB/1974

FOREWORD

The accident at the Fukushima Daiichi nuclear power plant in Japan caused by the seismically induced tsunami of 11 March 2011 involved core damage and radioactive release from three units and challenged the capability to maintain safety functions on additional units at the nuclear power plant and at other sites. The consequences and impact of the emergency for people and the environment underlined the need to assess the nuclear safety of multiunit sites against potential internal and external hazards and combinations thereof. The safety assessment of a single-unit site against internal and external hazards has been the focus of many methodology publications and standards. Prior to this accident, there were several known accident precursors that had highlighted the potential for multiple reactor accidents. The assessment of a multiunit site against multiple concurrent and correlated hazards has generally not received much attention in previous publications employing probabilistic safety assessment (PSA) methodology and has not been the subject of extensive studies by the nuclear industry or regulatory bodies.

IAEA Safety Standards Series No. SSG-3, Development and Application of Level 1 Probabilistic Safety Assessment for Nuclear Power Plants, published in 2010, provides general guidance for a Level 1 PSA and its application. However, it does not explicitly cover aspects of multiunit risk assessment. Currently, SSG-3 is being revised to address PSA considerations for multiunit sites. IAEA Safety Reports Series No. 96, Technical Approach to Probabilistic Safety Assessment for Multiple Reactor Units, published in 2019, summarizes the historical background of multiunit PSA (MUPSA) and offers insights from the review of the accident at the Fukushima Daiichi nuclear power plant and previous accident precursors, as well as providing a high level technical approach for identifying and assessing both internal and external event MUPSAs.

This publication provides technical details for analysing site specific internal and external hazards and the overall site risk. It presents a detailed methodology for developing a MUPSA model that has been tested via case study. The approach assumes that the licensee has performed a PSA for each unique reactor unit on the site to develop a MUPSA.

This safety report complements the IAEA safety standards, providing a technical basis for analysing the potential internal events and internal and external hazards, including combinations thereof, to be considered in a MUPSA.

The IAEA would like to thank the many PSA experts who participated in the consultancy and technical meetings for their valuable contributions to this publication, in particular D. Henneke (United States of America), P. Amico (United States of America), K. Fleming (United States of America), P. Hlaváč (Slovakia), A. Maioli (United States of America), P. Boneham (United Kingdom),

D. Kim (Republic of Korea), H. Jeon (Republic of Korea), W. S. Jung (Republic of Korea) and M. Jae (Republic of Korea).

The IAEA officers responsible for this publication were O. Coman and S. Poghosyan of the Division of Nuclear Installation Safety.

EDITORIAL NOTE

CONTENTS

1. INTRODUCTION

The use of probabilistic safety assessment (PSA) in supporting safety related decision making for nuclear power plants (NPPs) is most often based on the analysis of a single reactor unit; however, most operating NPPs have more than one reactor unit. According to the IAEA's Power Reactor Information System (PRIS) database, 139 out of 191 NPP sites are multi-unit (MU)[1] sites. Safety analyses and PSA are usually confined to a single unit (SU) with the assumption that other co-located units are safe, although some aspects of the dependency of the structures, systems and components (SSCs) of the analysed unit on co-located units are normally considered; therefore, the potential for accident sequences involving two or more reactor units, such as what occurred during the accident at the Fukushima Daiichi NPP, is not considered.

The PSA and safety analysis community have been aware of the potential for MU accidents for more than three decades, and this evidence is summarized in IAEA Safety Reports Series No. 96, Technical Approach to Probabilistic Safety Assessment for Multiple Reactor Units [1]. However, the accident at the Fukushima Daiichi NPP in 2011 reinforced the possibility of MU accidents, especially as a result of external hazards or a combination thereof. Section 2 of Ref. [1] provides a detailed discussion of the history of multi-unit probabilistic safety assessment (MUPSA), and of the lessons learned from the accident at the Fukushima Daiichi NPP and from the accident precursors that preceded it, with a focus on the MU risk and MUPSA. This background is not repeated here, other than to provide a reminder that the importance of MUPSA development has been known for many years but was brought to the forefront only as a result of this accident.

Determining the risk insights for MU sites using PSA requires explicitly addressing the interactions and dependencies among the units (both with positive and negative safety impacts) in a systematic and complete manner, including a delineation of SU and MU accident sequences. The MUPSA considers whether accident progression in one unit could induce or exacerbate events in other units in terms of preventing or mitigating multiple core damage events.

Determining the risk for an MU site is highly dependent on the scope, resolution and level of detail used for the supporting single-unit probabilistic safety assessment (SUPSA). This report focuses on the specific aspects of a MUPSA and on an explicit modelling approach. Alternative approaches that cover a large spectrum of SUPSA conditions are discussed in the annexes.

---

[1] Note that certain sites described in the IAEA PRIS database could be considered as one site from the MUPSA perspective.

1.1. BACKGROUND

Since 2012, the IAEA has initiated a number of activities intended to develop a framework and methods for the safety assessment of MU sites under the impact of multiple hazards. The following publications were developed as a result of those efforts:

— IAEA Safety Reports Series No. 96, Technical Approach to Probabilistic Safety Assessment for Multiple Reactor Units [1], covering the technical approach for Level 1 to Level 3 MUPSA;
— IAEA Safety Reports Series No. 92, Consideration of External Hazards in Probabilistic Safety Assessment for Single Unit and Multi-unit Nuclear Power Plants [2];
— IAEA-TECDOC-1804, Attributes of Full Scope Level 1 Probabilistic Safety Assessment (PSA) for Applications in Nuclear Power Plants [3].

In response to requests from Member States, in December 2016 the IAEA launched a new MUPSA project implemented by IAEA's Division of Nuclear Installation Safety (hereafter referred to as the MUPSA project). The project was supported by the technical basis developed in Refs [1, 3] and ongoing IAEA activities on risk aggregation [4] and supplemented by a state of the art review on MUPSA conducted by the IAEA [5].

The motivation for the MUPSA project was to develop a technical basis for providing guidance to meet the following safety requirements:

— Paragraph 5.15B of IAEA Safety Standards Series No. SSR-2/1 (Rev. 1), Safety of Nuclear Power Plants: Design [6], states: "For multiple unit plant sites, the design shall take due account of the potential for specific hazards to give rise to impacts on several or even all units on the site simultaneously."
— Paragraph 4.36A of IAEA Safety Standards Series No. GSR Part 4 (Rev. 1), Safety Assessment for Facilities and Activities [7], states: "For sites with multiple facilities or multiple activities, account shall be taken in the safety assessment of the effects of external events on all facilities and activities, including the possibility of concurrent events affecting different facilities and activities, and of the potential hazards presented by each facility or activity to the others."
— Paragraph 6.11 of IAEA Safety Standards Series No. GSR Part 7, Preparedness and Response for a Nuclear or Radiological Emergency [8], states: "For a site where multiple facilities in category I or II are collocated, an appropriate number of suitably qualified personnel shall be available to

manage an emergency response at all facilities if each of the facilities is under emergency conditions simultaneously."

IAEA resolution GC(62)/RES/6 [9], operative paragraph 58, states that the IAEA General Conference "[e]ncourages Member States that have not already done so to perform safety assessments, including at multi-unit sites, to evaluate the robustness of nuclear power plants and other installations against multiple extreme events, and share their experience and the results of such assessments with other interested Member States". In addition, IAEA resolution GC(61)/RES/8 [10], operative paragraph 60, states that the IAEA General Conference "[r]equests the Secretariat to continue efforts to develop guidance on the safety of multi-unit sites".

The MUPSA project provides practical details on how to implement MUPSA. For pragmatic reasons, it is mainly focused on Level 1 MUPSA. The MUPSA project was implemented in three phases:

— Phase I — develop a document providing a methodology for the implementation of MUPSA, with practical PSA modelling guidance that complements the high level technical approach to MUPSA detailed in IAEA Safety Reports Series No. 96, Technical Approach to Probabilistic Safety Assessment for Multiple Reactor Units [1];
— Phase II — develop a case study following the methodology developed in Phase I [11];
— Phase III — improve the methodology based on the lessons learned from the case study developed in Phase II and integrate the improved methodology and the case study into a single document (this safety report).

The Phase II case study was an essential part of the MUPSA project, intended to complete the efforts implemented in the project's Phase I and to support the finalization of the project in Phase III.

In addition, aspects related to risk aggregation in the MU context are covered in the IAEA project on Development of a Methodology for Aggregation of Various Risk Contributors for Nuclear Facilities [4], which is completed and was used to support the MUPSA project.

The above mentioned references provide high level guidance for performing a MUPSA, practical guidance on modelling details and the results of a case study applying the MUPSA methodology. This safety report presents an improved detailed methodology on the basis of the feedback from the case study and is the outcome of the MUPSA project.

1.2. OBJECTIVE

The main objective of this publication is to present approaches for conducting a MUPSA for a new or existing NPP. The publication provides updated information on different aspects relating to estimating the MU risk, evaluating initiating events (IEs) induced by internal and external event hazards and evaluating factors affecting MU risk.

This publication is intended for use by operating organizations, designers, PSA practitioners, regulatory bodies and researchers. It provides a technical basis for MUPSA, in accordance with the IAEA safety standards. It can be used to develop guidelines for conducting MUPSA in relation to internal and external events.

Guidance provided here, describing good practices, represents expert opinion but does not constitute recommendations made on the basis of a consensus of Member States.

1.3. SCOPE

The scope of this publication covers the major tasks of MUPSA development. It describes the general MUPSA approach as well as the details of the MUPSA case study implemented within the MUPSA project. The publication provides a detailed description of Level 1 MUPSA methodology and experiences available in Member States. In addition, it outlines the principles of development of the Level 2 MUPSA model and the path forward for MU consequence analysis (e.g. Level 3 MUPSA).

The scope of this safety report addresses the consideration of various hazards and plant operational states (POSs) normally considered in PSA development in the MU context. It also covers the integration of SUPSA with MUPSA results to obtain the complete risk profile for the site (see discussion in Section 4.5.1).

The hazards arising from malicious acts are not outside the scope of this safety report.

1.4. STRUCTURE

This safety report comprises of six sections, three appendices and six annexes. Section 2 focuses on the technical basis for the MUPSA methodology, including international experience in MUPSA. Section 3 documents the assumptions used in the methodology. The methodology is detailed in Section

4, providing a step-by-step process for performing MUPSA. Section 5 discusses the Level 2 and Level 3 modelling considerations, including Level 2 interface considerations with the Level 1 PSA. Finally, Section 6 provides a summary of the methodology and a path forward.

The appendices provide supplemental information. Appendix I provides a summary of the level of changes necessary to perform the MUPSA using the SUPSA model. Appendix II summarizes the case study results and insights. Appendix III provides supplemental details on approaches used to model the MU accident sequence analysis.

The approaches and experiences of Member States (Canada, France, Hungary, the Republic of Korea, the United Kingdom (UK) and the United States of America (USA)) related to MUPSA are presented in the annexes.

2. OVERVIEW OF MUPSA TECHNICAL BASIS

2.1. LIMITATIONS AND CHALLENGES

A broad overview of the technical challenges of MUPSA is well described in Safety Reports Series No. 96 [1], in the summary of the large international workshop held in Canada in 2014 [12] and in the Organisation for Economic Co-Operation and Development (OECD)/Nuclear Energy Agency (NEA) Working Group on Risk Assessment task report, Status of Site-Level (Including Multi-Unit) Probabilistic Safety Assessment Developments [13]. These and other challenges were also indicated during the MUPSA state of the art review performed by the IAEA and in particular its Division of Nuclear Installation Safety (NSNI) before initiating the development of MUPSA methodology [10].

A summary of the limitations and challenges of MUPSA, including areas needing further development, is as follows:

(1) There is limited industry experience and practical guidance on performing a MUPSA and on the use of SUPSA results for planning and developing a MUPSA.
(2) Limited guidance is available on the identification and screening of IEs for MUPSA purposes.
(3) Strategies are needed for managing the modelling complexity of different combinations of initial POSs from which to build a PSA model.

(4) The impact of radiological contamination of the site from one unit on the equipment operability, operator actions and accident management measures of the other units on-site needs to be taken into consideration.

(5) Definition of the risk metrics that need to be adapted from SUPSA to resolve the different end states of a MUPSA and applied consistently within the context of a MUPSA is required.

(6) There are technical issues with aggregating risk contributions from accident sequences involving different hazards and end states.

(7) Common cause failure (CCF) models and supporting data analysis need to address groups of components that can be in different reactor units.

(8) Modelling of interunit dependencies such as shared systems and areas with shared hazards need to be addressed in the MUPSA.

(9) Human reliability analysis (HRA) needs to consider aspects of the MUPSA (e.g. procedures, dependencies, decision making) that cannot be included in the SUPSA or can involve new performance shaping factors associated with a MU accident.

(10) The MUPSA needs to consider how to construct a MU risk model from existing SUPSA models, combining accident sequences for multiple units (potentially a large number of units) and developing new models that focus on MU accidents, including the treatment of internal and external hazards and the modelled fragilities.

(11) The MUPSA includes new end states involving MU accidents, including Level 2 release categories (RCs). There can be challenges in the interpretation of PSA results, particularly with respect to Level 2 and Level 3 PSA metrics.

(12) Interunit correlation of SSC fragilities to external hazards, such as a seismic event, is a challenge to model in the MUPSA.

(13) There is currently no method to provide correlation of many of the factors used in calculating Level 2 MU risk. Phenomenological factors and their correlations need to be addressed and understood in MU risk, as they can significantly affect the MUPSA results.

(14) Risk aggregation for MU accident sequences introduces new factors such as MUPSA assumptions, scope and modelling simplifications, some of which depend on the MUPSA accident sequence modelling approach used. The MUPSA method needs to consider these factors, including simplifications used in the MUPSA model development.

Regarding item 1, industry experience continues to grow and lessons learned are discussed in Section 2.2.

Items 2–11 and 14 are addressed within Section 4 of this report based on the actual state of practice. Item 12 is discussed in Section 4.4.5, but technical advancements continue in this area; in particular, in terms of the practical application of methodologies for fragility correlation to MUPSA models. Finally, item 13 is discussed in Section 5.

2.2. OVERVIEW OF AVAILABLE EXPERIENCE

The following sections provide a summary of the references used in the development of the MUPSA methodology. Section 2.2.1 includes the previous IAEA experience prior to development of the IAEA methodology and the case study, discussed in Section 2.3. Section 2.2.2 provides a high level summary of other international experience. Section 2.2.3 discusses whether scoping approaches have been found to be suitable for MUPSA risk analysis.

2.2.1. Previous IAEA experience

As described above, the MUPSA project was built on the technical basis available in Member States using the relevant IAEA publications. A state of the art review performed by NSNI/IAEA and discussions during the consultancy meetings identified the MUPSA technical basis for this report, which includes the following IAEA publications:

— IAEA Safety Reports Series No. 92, Consideration of External Hazards in Probabilistic Safety Assessment for Single Unit and Multi-unit Nuclear Power Plants [2];
— IAEA Safety Reports Series No. 96, Technical Approach to Probabilistic Safety Assessment for Multiple Reactor Units [1];
— IAEA-TECDOC-1804, Attributes of Full Scope Level 1 Probabilistic Safety Assessment (PSA) for Applications in Nuclear Power Plants [3].

Additional resources available to support this effort include the preliminary results of the completed IAEA project on Development of a Methodology for Aggregation of Various Risk Contributors for Nuclear Facilities [4]. In the context of multisource considerations, the risk aggregation assumes the combined representation of risks coming from different sources of radioactivity available at the site (e.g. reactor cores, spent fuel pools (SFPs), dry spent fuel storages and other facilities with radioactive sources).

An IAEA Technical Meeting on MUPSA (held in October 2019) included presentations by various Member States on MUPSA activities and detailed technical discussions. The following provides a summary of insights from the technical meeting:

(a) Most sites with multiple units showed that MU risk, such as multi-unit core damage frequency (MUCDF), can represent a high percentage of SU risk. However, the contribution of MU risk to the overall site risk varied depending on both the unit designs and the site:

    (i) For highly independent units of similar design, this risk was dominated by external hazards;

    (ii) For reactors with shared systems, their failure can represent a significant fraction of MU risk;

    (iii) For sites with relatively low risk from external hazards, MU risk can make a much lower contribution to overall site risk;

    (iv) Specific designs can result in higher MUPSA risk contribution from internal hazards, even for units without significantly shared systems (e.g. high interunit CCF probabilities could result in a higher MUCDF).

(b) For some sites, MU CCF needs to be considered carefully to properly characterize its risk contribution. However, there is not much available data to support detailed evaluations of MU CCF. This is discussed further in Section 6.1, which focuses on the path forward.

(c) MUPSA applications indicate that the existing HRA methods are applicable and functional for MUPSA. However, the MUPSA includes unique performance shaping factors and dependency models that need to be considered. This is discussed further in Section 4.4.4, although the development of updated guidance on HRA is not provided.

(d) Human and organizational factors were also noted as impacting on MU HRA, as was noted in the lessons learned from the accident at the Fukushima Daiichi NPP. Consideration of human and organizational factors in existing HRA methods is also a challenge for SUPSA, which becomes more complex in the MU context. This topic is discussed in the forthcoming IAEA safety report on human reliability analysis for nuclear facilities, which provides information on the extent to which human and organizational factors are considered in current HRA methods [14].

(e) Where site level safety goals were developed, they were focused on releases from several units and their consequences. Therefore, Level 2 and Level 3 MUPSA goals were more likely to be needed at the site level, such as site releases or dose levels. Most participants in the technical meeting who were considering safety goals noted that there was no need to develop site level core damage frequency (CDF) safety goals. It was commonly accepted that risk metrics on site level CDF (MUCDF or site core damage frequency (SCDF)) are to be applied for the ranking of various risk contributors, but not for comparison against safety goals.

2.2.2. Previous international experience

MUPSA concepts have been investigated for many years, starting with the Seabrook Level 3 assessment, which included MUPSA considerations and was completed in 1983 [15]. More recent studies include the following:

— ASME/ANS: Non-mandatory Appendix Probabilistic Risk Assessment of Multi-Unit Plants and Sites[2] [16] and MUPSA requirements extracted from the US Probabilistic Risk Assessment Standard for Advanced Non-LWR Nuclear Power Plants [17];
— Canadian Nuclear Safety Commission (CNSC): Summary Report of the International Workshop on Multi-Unit Probabilistic Safety Assessment [12];
— CANDU Owners Group: Development of a Whole-Site PSA Methodology [18];
— European Commission: Deliverable reports of the European Advanced Safety Assessment Methodologies: Extended PSA (ASAMPSA_E project) [19–23];
— Electric Power Research Institute (EPRI): Publications on risk aggregation [24, 25], MUPSA [26] and the MUPSA framework [27];
— GE Hitachi: Methodology of MUPSA for advanced boiling water reactors (ABWRs) [28];
— OECD/NEA: ICDE Topical Report on Collection and Analysis of Multi-Unit Common-Cause Failure Events [29];
— Korea Atomic Energy Research Institute (KAERI): A methodology and software package for MUPSA with the results of their application to a six-unit NPP site [30–34] and a pragmatic approach to modelling CCFs in MUPSA [35, 36];
— Korea Hydro & Nuclear Power (KHNP): MUPSA project for the Kori/Saeul NPP site with nine reactor units and its preliminary results [30–34, 37, 38];
— OECD/NEA/Committee on the Safety of Nuclear Installations: Status of Site-Level (Including Multi-Unit) PSA Developments [13];
— Oak Ridge National Laboratory: Study on IE analysis for multi-reactor plant sites [39];
— UK Office for Nuclear Regulation (ONR): Safety Assessment Principles (SAPs) for MU sites [40] and Level 3 MUPSA assessment [41];
— US Nuclear Regulatory Commission (NRC): Technical approach on integrated site risk analysis task in the Level 3 PRA project [42];

---

[2] The non-mandatory appendix was issued in draft form and was never approved for use by ASME/ANS. However, the non-mandatory appendix was used in developing the non-LWR standard, which was issued for trial use.

— Various papers from international conferences [43–45].

Detailed examples of available international experience in this area are provided in the annexes of this publication.

2.2.3. Experience with scoping approaches

It is not clear that a full quantification of MU risk is needed for all applications. If the analysis of MU risk is performed in order to only get the total risk estimates/metrics, such as site CDF, and a conservative estimate of these metrics is acceptable, a scoping approach may be sufficient for this application [46, 47]. One scoping approach is described in Section VI–3 of Annex VI, which was applied to the ABWR PSA in the UK.

If the analysis of MU risk is initially estimated using the scoping approach and the results meet the goal for the site, it may be that a detailed analysis is not needed. For example, if the MUPSA is being performed to calculate a site CDF estimate and the resulting scoping analysis is below the site CDF goal, then detailed analysis is not necessary. In this case, the scoping approach needs to demonstrate that the results are conservative and consider interactions between all units at the site. However, the results of the scoping approach do not support applications where the results of risk measures or similar are needed.

2.3. RISK METRICS

As concluded in Refs [1, 10], there is a need for additional risk metrics beyond those used for SUs (e.g. CDF, large early release frequency (LERF)) to fully express the risk profile in the MU context. Safety Reports Series No. 96 [1] and the ongoing IAEA project on risk aggregation [4] describe the risk metrics to address the MU risk for Level 1, Level 2 and Level 3, which have been expanded beyond those identified in Ref. [1].

The IAEA Phase I and II work focused on the risk metrics associated with core damage (fuel located in the core). However, in the longer term, MUPSA will need to consider fuel or sources located in areas other than the core.

When considering MU fuel damage in the SFP, other terminology is used to represent combinations of fuel or core damage at multiple units. Figure 1 presents an example of a visual interpretation of potential radioactive sources on a nuclear plant site and their taxonomy used in this report. As shown in Fig. 1, the plant site could potentially include reactors, SFPs (both types — individual for each reactor and common SFPs), dry spent fuel storages and other facilities with radioactive sources that could lead to releases. For the purpose of this report, the unit is defined as the reactor with its individual SFP, or just the reactor itself if an individual SFP is not available. Dry spent fuel storages, common SFPs and other facilities are considered as different radioactive sources that could lead to releases. Each block shown in Fig. 1 (e.g. reactor, individual SFP, facility F1) is considered as a radioactive source.

As mentioned in Ref. [1], the SU risk metrics (e.g. single-unit core damage frequency) are not the same as the risk metrics traditionally used in PSA. The traditionally used risk metrics (e.g. CDF, large release frequency (LRF), LERF) are normally expressed in units of events per reactor-year, are associated with each reactor on the site and are calculated from the SUPSA. The above MUPSA metrics are expressed in units of frequency per site-year. This change is needed to provide a coherent basis for aggregating risk contributions from accident sequences involving different end states. The following sections provide a summary of site, SU and MU Level 1, 2 and 3 risk metrics.

2.3.1. Multi-unit Level 1 risk metrics

Reference [1] provides various risk metrics to complement the traditional PSA metrics. The risk metrics presented therein for Level 1 MUPSA are mainly related to the reactor core.

There are two categories of core damage events that are included in the estimation of SCDF: those events that involve core damage on a SU and those that involve core damage on multiple units [1]. The following list presents the definitions of core damage related risk metrics:

— **Single-unit core damage frequency (SUCDF):** frequency per site-year of an accident involving core damage on 'one and only one reactor' on a MU site. SUCDF is the aggregated frequency of core damage that accounts for

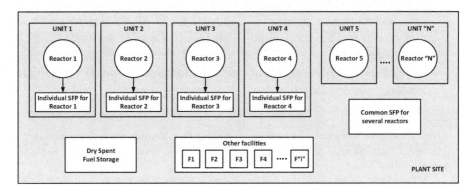

FIG. 1. Visual interpretation of radioactive sources on a plant site.

all the reactors on the site but includes accident sequences involving core damage only on a SU.

— **Multi-unit core damage frequency (MUCDF):** frequency per site-year of an accident involving core damage on two or more reactors on a MU site.
— **Site core damage frequency (SCDF):** frequency per site-year of an accident involving core damage on one or more reactors.[3]

As presented in Fig. 1, in addition to the reactors, site level risk assessment could include other sources of radioactivity, such as SFPs and dry spent fuel storage facilities. Considering these, alternate metrics may be developed for Level 1 MUPSA considering fuel damage on single or multiple sources (e.g. reactors, SFP). Examples of such definitions are:

— **Multisource fuel damage frequency (MSFDF):** the frequency per site-year of an accident involving fuel damage from two or more sources on a MU site;
— **Site fuel damage frequency (SFDF):** the frequency per site-year of an accident involving fuel damage on one or more sources on a MU site.[4]

2.3.2. Multi-unit Level 2 risk metrics

The risk metrics presented in Ref. [1] for Level 2 and Level 3 MUPSA are described for a site level and consider reactors as well as other facilities available on-site:

— **Site large early release frequency (SLERF):** the frequency per site-year of an accident involving a large early release, either from one or more reactors or on-site facilities.
— **Site release category frequency (SRCF):** the frequency per site-year of each distinct release category (RC) for a Level 2 MUPSA given a release from one or more reactors or on-site facilities. These RCs include the RCs already defined in a Level 2 SUPSA for each reactor or on-site facility and for releases from a SU, as well as categories for accidents involving multiple reactors or on-site facilities.

---

[3] These definitions do not include contributions from non-reactor sources such as the SFP or other installations. These are addressed further in the discussion involving SFDF or similar.

[4] For SFDF, the estimate does not include contributions from fuel in the reactor cores. When estimating MU fuel damage frequency or similar, the metric would include fuel damage in any location including in the reactor cores.

The above mentioned risk metrics are defined for a site level and consider both single and multiple reactors or on-site facilities. They are applicable regardless of the source taken into consideration (e.g. core or SFP). Depending on the objectives of the risk assessment, alternate metrics may be developed for Level 2 and Level 3 MUPSA, considering undesired end states only on multiple reactors or on-site facilities. Examples of such definitions are:

— **Multi-unit large early release frequency (MULERF):** the frequency per site-year of an accident involving large early release on two or more reactors or on-site facility on a MU site;
— **Multi-unit release category frequency (MURCF):** the frequency per site-year of an accident involving a RC involving two or more units on a MU site;
— **Multisource release category frequency (MSRCF):** the frequency per site-year of an accident involving a RC involving two or more sources on a site.

2.3.3. Multi-unit Level 3 risk metrics

Public risk measured under a Level 3 PSA takes into account a number of possible risk measures, including effective dose, individual risk and societal risk, such as economic consequences. Given that the Level 2 MUPSA risk metrics are calculated including MU RC frequencies, the selected Level 3 risk metrics can be calculated for the site in a similar manner as the traditional SU Level 3 risk metrics. In fact, in the Level 3 domain the risk measures for SU and MU are no different but would simply be identified as the effective dose from a SU or for the site.

Currently there is no international consensus related to safety goals for MU sites. If a site level safety goal is applicable, the methodology below can be used to calculate the frequency for the applicable risk measure that can be used to measure the site goal. If no site safety goal is applicable, the MUPSA can be used to develop insights that go beyond the SUPSA insights, as discussed in Section 4.6.

Further discussion on safety goals is provided in IAEA-TECDOC-1874, Hierarchical Structure of Safety Goals for Nuclear Installations [48]. It provides a high level framework and examples on establishing a consistent and coherent hierarchical set of safety goals for nuclear installations.

Concluding the discussion on risk metrics, it needs to be stressed that the examples provided above are not exhaustive, and other risk metrics might be defined and used, based on the objectives of risk assessment. Also, other risk metrics might be developed for shutdown POSs, if the SUPSA analyses separate metrics (such as boiling or fuel uncovery).

2.4. IAEA/NSNI PROJECT ON MUPSA

As discussed in Section 1.1, the implementation of the MUPSA project comprises the following phases:

— Phase I — develop a document providing a methodology for the implementation of MUPSA with practical PSA modelling guidance that complements the high level technical approach to MUPSA found in Safety Reports Series No. 96 [1];
— Phase II — develop a case study following the methodology developed in Phase I [11];
— Phase III — improve the methodology based on the lessons learned from the case study developed in Phase II and integrate the improved methodology and the case study into a single document (this safety report).

The details of Phase I and Phase II are provided in Sections 2.4.1 and 2.4.2, respectively.

2.4.1. Phase I. MUPSA methodology

The Phase I methodology was based on the experience discussed in Section 2.2, using the technical approach in Ref. [1] as a starting point. Phase I of the current project resulted in a working material document, providing a methodology for the implementation of MUPSA with practical PSA modelling tips. One of the key assumptions of the MUPSA methodology developed in Phase I is that MUPSA needs to be developed based on SUPSA models of sufficient quality, which are to be built on the basis of the current state of practice on PSA. In this regard, the MUPSA methodology assumes implementation of the typical PSA tasks with proper consideration of the MU context. The level of changes necessary for typical MUPSA tasks in comparison with traditional SUPSA practices was evaluated based on the engineering judgement in Phase I of the project and was later revised based on the experience gained during the case study results (see more detailed information in Appendix I).

The refinements to the methodology based on the Phase II case study are discussed in Section 2.4.2 and the updated methodology is provided in Section 4.

2.4.2. Phase II. MUPSA case study

The objective of the case study in Phase II was to verify the proposed Phase I MUPSA methodology by applying it to a realistic NPP configuration using a realistic PSA model and to provide feedback on the applicability of the proposed methodology for standard PSA tasks. In addition, the case study provided a base for improvement and increased the level of detail reflected in the methodology.

The specification of the case study was developed and agreed on by the IAEA/NSNI MUPSA working group and has the following details:

— The case study was designed in such a way as to deal with and verify various aspects of the MUPSA methodology.
— The case study reflected the potential complexity of the MUPSA task, depending on the number of units available at a typical NPP site, their type and configuration, while considering a fixed number of units (four). Sites with more units will have further complexities, while each site will have unique complexities.
— The case study covered various categories of multi-unit initiating events (MUIEs) addressing the wider variety of typical accident scenarios. The case study did not exercise all possible hazards impacting on the site but was extended to sequences beyond internal events, including seismic and fire sequences.

The case study provided a better understanding of the level of effort necessary to expand current PSA studies to the MU context and identify the main challenges for MUPSA. The case study results were not intended to be extrapolated beyond the feasibility of the approach, such as applicability to a specific site.

The case study demonstrated how the general principles of PSA need to be applied in a MU context. Thus, realistic modelling is important to obtain a sufficient basis for risk informed decision making. In this context, it is necessary to highlight that both positive and negative interactions of units have been considered, which allowed analysts to obtain a realistic risk profile as an output of the case study.

The case study provided feedback on the applicability of the proposed methodology for standard PSA tasks, concluding that approaches presented in Appendix III are applicable for the construction of MU event trees and lead to the same results.

The Phase I methodology identified during the Phase II case study review includes the following limitations:

(a)   The risk metrics used in the case study was limited to Level 1 MUPSA.
(b)   The screening process, when applied to the IE selection, including mode/POS combinations, has not been fully tested. The process of IE selection and screening can be implemented in accordance with the described process, but the potential contribution from the screened out risk has not been evaluated.
(c)   Similarly, the screening of low risk sequences from the original SUPSA, used to reduce the size of the MUPSA, has not been fully tested. Again, the potential cumulative contribution from screened out sequences has not been evaluated.
(d)   The case study effort in Phase II modelled MUIEs for different hazard groups but did not model external floods and high winds.
(e)   The case study effort in Phase II exercised the steps in the methodology, with the following exceptions:
   (i)   The case study did not exercise the screening steps for accident sequences;
   (ii)   The case study did not analyse partial correlations of seismic fragilities (only full fragility correlation was assumed);
   (iii)   The case study did not review external hazard induced failures (e.g. seismic induced fire) of SSCs that were screened out from the SUPSA to ensure that this screening remains valid in the MUPSA;
   (iv)   The case study did not fully complete the interpretation of results, as discussed in Section 4.6.

A summary of the Phase II methodology and results is provided in Appendix II. The MUPSA Phase II case study above provided a number of suggested improvements to the methodology. The suggested improvements are discussed in Appendix II, Section II.5.

3. MUPSA METHODOLOGY: PRINCIPLES AND ASSUMPTIONS

Because there are many configurations of MU sites in the various Member States, and additional configurations are expected to arise in the future, the

MUPSA methodology is intended to be flexible and practical. The guiding principles of the development of the MUPSA methodology include the following:

— Be in line with IAEA Safety Reports Series No. 96, Technical Approach to Probabilistic Safety Assessment for Multiple Reactor Units [1]; SSG-3, Development and Application of Level 1 Probabilistic Safety Assessment for Nuclear Power Plants [49]; SSG-4, Development and Application of Level 2 Probabilistic Safety Assessment for Nuclear Power Plants [50]; and NS-G-2.13, Evaluation of Seismic Safety for Existing Nuclear Installations [51] and consider the MUPSA attributes in IAEA-TECDOC-1804 [3];
— Be applicable to all modelled POSs (e.g. full power operation versus low power/shutdown mode) and different combinations of modes for each reactor unit (screening of combinations of modes/POSs is included in the methodology);
— Address internal events and internal and external hazards with the PSA state of practice (screening of hazards is included in the methodology);
— Be functional for sites with many units;
— Be functional for sites with different types of units;
— Be able to identify the risk (significant contributors to MU risk) and provide realistic quantitative estimates of that risk and associated uncertainties consistent with the selected MUPSA risk metrics;
— Be implementable with state of practice resources and not be overly burdensome when compared with SUPSA modelling.

Finally, the methodology does not seek to solve technical issues in PSA that still exist in SUPSA. That is, if a method or approach to the PSA represents a technical challenge to implement in SUPSA, it is not suggested to go beyond the current state of practice.[5]

The general assumptions applied in the methodology, including references to specific sections representing the technical basis and more detailed information on the topic are described in the following list:

(1) SU models representative of each unit on the site are available with sufficient quality (i.e. in accordance with the IAEA safety standards), and the scope of

---

[5] The MUPSA modelling is intended to extend SUPSA methods into the MUPSA environment. For example, it is not the intent to suggest that the MUPSA needs to implement a new methodology for CCF analysis, HRA or seismic correlation; but rather, to implement the current methodology used in the SUPSA and to offer suggestions on how to cover the respective MU context.

those models (e.g. in terms of the hazards considered) is representative of the expected scope of the MUPSA.

(2) Administrative shutdowns of otherwise unaffected units could be screened out based on low risk significance when considering MU impacts (see Section 4.3.1). This screening does not include administrative shutdowns before the event (such as external flooding or hurricane), which unit impact can occur after the shutdown.

(3) IEs that affect only a SU can be excluded from a model that is used to represent MU core damage accident sequences (see Section 4.3.1), while exclusively SU sequences may be retained and incorporated into an integrated risk model that address both single and MU events. However, some additional consideration for consequential loss of off-site power (LOOP) might be needed (see Sections 4.3.1.1 and 4.2.1.3).

(4) SU core damage sequences in the MU model do not need to be considered, unless they lead to a MU sequence. These sequences are evaluated using the existing SU model and the results are combined with the MUPSA model results after quantification.

(5) Interunit CCF would generally not apply when the units are of different type, model or vintage. However, it needs to be verified that component groups (e.g. diesel generators) are indeed different between the units, given that even different unit vintages can have SSCs of the same manufacturer, characteristics, etc. (see Section 4.4.5.2).

(6) Interunit CCF (non-hazard correlations) generally does not apply to passive components, except in cases where the passive components' reliability is significantly impacted by common factors. IEs failing passive components, such as MU heat exchanger or strainer plugging when MU systems draw from the same raw water source, could need to be considered if not otherwise screened out (e.g. based on low risk contribution) and not modelled as a MUIE.

(7) There is generally no need to expand the correlation groups beyond the SUPSA correlation groups, other than for single components in a unit with identical components in another unit (e.g. turbine driven auxiliary feedwater pumps), which are added to the MUPSA CCF model (see Section 4.4.5.2). Any new CCF component groups need to be considered for new correlation groups (see Section 4.4.5.3).

(8) Low risk scenarios in the SUPSA can be screened out from the MUPSA modelling. This includes low risk IEs as well as low risk accident sequences for retained IEs. The screening process needs to consider both the impact on the SUPSA and the potential MU impact (see Section 4.2). Exceptions are discussed below in Assumption 9.

(9) Hazard-induced failures of individual SSCs that are screened out from the SUPSA for a given hazard can generally be screened out from the MUPSA (i.e. the SSC screening criteria for each hazard used in the SUPSA are still valid for the MUPSA; see Section 4.2). However, if screening from a SUPSA is based on the frequency of accident sequence where shared equipment across units is credited in the frequency estimate, the MU risk could need to be evaluated (see Section 4.3.1.3).

(10) Interunit seismic correlation failure does not apply when the seismic input to identical SSC is significantly different (see Section 4.4.5.3).

(11) For screening HRA purposes, in the case of operator actions where staff are shared between units, the shared personnel are credited to the second unit in the sequential modelling unless operating procedures and operators' training direct otherwise (see Section 4.4.4).

(12) For screening HRA purposes, where one or more units are in core damage, the HRA modelling assumes the worst-case core damage progression (greatest challenge to the operators at each unit) of the unit(s) in core damage. For example, if an operator action outside the main control room (MCR) is potentially affected by dose release from the other unit, the screening human error probability (HEP) is based on the assumption that core damage on the other unit has occurred before the operator action is performed. This could translate into not crediting any local actions when performing screening HRA, depending on the dose release and timing.

(13) The MUPSA can be simplified to model representative core damage combinations to limit the size and complexity. Similarly, it is not necessary to perform the quantification for all possible unit combinations or POS combinations. This is not to say that other permutations or other POS combinations are not possible. Rather, assessing a single permutation of a given order (e.g. two-unit core damage, three-unit core damage) and in a given sequential order as representative of the set of permutations is sufficient, and the 'missing' combinations can be added to the results to get a reasonable estimate of the total frequency of all unscreened permutations of a given order (see Section 4.2.1.3 and Appendix III). It may be that the order of the combinations can affect the results, and the selected represented combination may be conservative. However, any simplification needs to be accounted for during the interpretation of the results or analysis of the importance measures.

(14) Type A human failure events (HFEs) identified for systems not common to multiple units are assumed to be independent.

(15) Development of a Level 2 MUPSA can be performed using similar steps as developing the Level 1 MUPSA, but with consideration for Level 2 specific factors, such as phenomenological events (see Section 5).

(16) A MUIE is assumed to not occur (or does not have to be modelled in the MUPSA) if a single-unit initiating event (SUIE) or hazard degrades a standby system at the second unit, but that unit can continue operating (see Section 4.3.1).

4. GENERAL APPROACH FOR LEVEL 1 MUPSA

The methodology in this publication expands on aspects discussed in Safety Reports Series No. 96 [1] that are relevant to the performance of a Level 1 MUPSA.

Figure 2 is a flow chart for the Level 1 MUPSA methodology presented in this report, with the section numbers that provide the details of each step noted. The detailed discussion of each step in the methodology is included in Sections 4.1–4.7. Although the process shown in Fig. 2 is visualized as linear, the process performed for the MUPSA is iterative. For example, if the initial results are greatly impacted by the application of conservative CCF or HRA, refinement of the CCF or HRA may be needed, followed by reperformance of subsequent steps.

4.1. LEVEL 1 MUPSA SCOPE AND RISK METRICS SELECTION

The scope of the Level 1 MUPSA is limited by the scope of the SUPSA and the intended applications of the MUPSA. That is, it is not the intent of this methodology to require that additional scope in terms of POSs and hazards be added to the SU model beyond those that have already been included. For example, if the SUPSA includes a full power and shutdown analysis for internal events, but only a seismic PSA for full power operations; it is not expected that a shutdown seismic PSA be completed in support of the MUPSA. Thus, the steps in the methodology assume that a SUPSA suitable for the development of individual base models as needed for the MUPSA is already available.

4.1.1. Scope selection

The first step in developing the MUPSA model is to determine what scope of the SUPSA will be included in the MUPSA, including any internal or external hazards, or the analysed POS. The scope needs to include all risk significant hazards and POSs included in the SUPSA, with the potential for MU impacts. The selected scope will later be reviewed for potential screening. Therefore, the initial scope determination does not include screening (e.g. screening of low risk contributors or potentially low risk POS combinations). For example, if refuelling operations is included in the shutdown SUPSA, but refuelling of multiple units

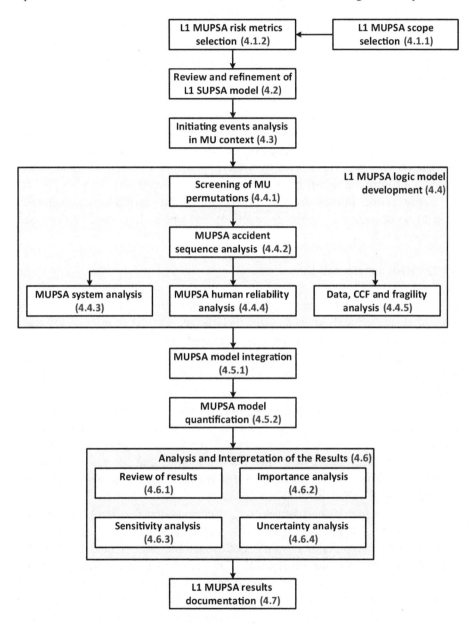

FIG. 2. *Level 1 (L1) MUPSA methodology. The relevant paragraphs of Section 4 are highlighted in blue.*

is not expected to occur, the POS combination for multiple units in refuelling would be included with the scope of the MUPSA but later screened as unlikely. Screening is discussed in Section 4.2.1.

The scope selected can impact the applicable risk metrics selected for the analysis, as discussed below, and documented as discussed in Section 4.7. For example, if the SFP PSA uses a risk metric of fuel damage frequency, the MUPSA risk metrics need to consider risk metrics involving MU fuel damage frequency as well as MUCDF.

The scope of both the SUPSA and MUPSA may need to be expanded, depending on the application of the MUPSA. For example, if the purpose is to verify that site safety goals have been satisfied, a more complete scope could be needed. In any event, the scope of the MUPSA could be controlled via the application of screening criteria by which parts of the PSA that, if included, would not contribute significantly to risk may be excluded.

Any scope limitations that could affect the MUPSA risk estimates need to be documented in the assumptions and uncertainty step, as discussed in Section 4.6.4.

4.1.2. Risk metrics selection

Several risk metrics are defined to complement the traditional PSA metrics that are associated with SUPSAs. This includes the site risk metrics and SU and MU risk metrics. These risk metrics are discussed in detail in Section 2.3 and are used to supplement the traditional reactor based risk metrics of CDF and LRF, which would continue to be used for individual PSAs that are performed for one reactor at a time.

The selection of the applicable MU risk metrics is determined based on the scope discussed in Section 4.1.1, as applied to the site and unit specific risk metrics discussed above. If the scope of the SUPSA involves only CDF, including damage to the reactor core only, then the risk metrics that address core damage on one or more units would be selected. These base Level 1 risk metrics typically include SCDF, SUCDF and MUCDF.

Other risk metrics may be selected, as discussed in Section 2.3, depending on the use of the MUPSA and whether the site includes more than two units. For example, if the site contained two pairs of two similar units (four units total), the Level 1 risk metrics could involve MUCDF for the two sets of two-unit pairs, plus a MUCDF for all four units. Additional risk metrics would be selected if the SUPSA involved Level 2 analysis and if the MUPSA included shutdown or SFP risk. The MUPSA Level 2 risk metrics are discussed further in Section 5.1.2. Finally, the results would also likely involve the calculation of the ratio of MU to

SU risk metrics. This MU ratio was calculated in the IAEA case study discussed in Appendix II.

The basis for the selection of risk metrics needs to be documented, as indicated in Section 4.7.

4.2. REVIEW AND REFINEMENT OF LEVEL 1 SUPSA MODEL

One of the key steps in the MUPSA model is to create PSA models for each unit on the site. If there are not individual PSA models for each unit, the steps involved (as discussed in Section 4.3) include duplicating the SUPSA model and modifying the models to account for MU aspects. This includes, for example, changing basic event (BE) and gate names for each unit, creating unit specific models, reviewing and correcting simplifications that affect the MUPSA model, modelling correctly shared equipment and cross-ties and reviewing undeveloped events for MU modelling aspects.

As mentioned in Section 3, the basic assumption is that SU models representative of each unit on the site are available in sufficient quality and with the required scope. Once the SU models are complete, each is analysed separately to ensure the results for SUCDF and single-unit large release frequency (SULRF) appear accurate before incorporating each into the MUPSA model. The MUPSA model involves accounting for the interunit dependencies, correlations and shared equipment accounting for the IE specific aspects, as discussed in detail in various subsections of Section 4.4.5.2.

Often SUPSA models apply the optimistic assumption that certain initiators might affect only the SU, and all site resources (material and human) are available for that unit. Also, any possible help from the other site units is often considered as being available for analysed unit. In reality, for accident conditions affecting more than the SU, these provisions may be not available. Finally, when modelling the Level 2 and Level 3 aspects of a SUPSA, the possibility that the other unit could also be experiencing a CDF event needs to be considered and modelled.

As a prerequisite to developing a MUPSA, accurate modelling of the site context in the SUPSA (e.g. internal events and hazards) is necessary, which involves the appropriate identification of dependencies between units, including:

— Shared SSCs (e.g. external power, pumping station, common buildings, support systems), including cross-ties that could be used to connect systems between units;
— Shared resources (e.g. water, fuel);

— Common site mitigation provisions (e.g. the site swing diesel, which can be used to support different units, and the capacity to support single or multiple units);
— Potential interunit CCFs (e.g. due to identical components, identical maintenance);
— Potential hazard correlations (e.g. seismic hazards, tsunami);
— Proximity dependencies, including shared MCRs;
— Human and organizational dependencies (e.g. shared MCRs, sharing and limitations of human resources, availability or lack of accident management procedures that can support a MU accident);
— Possibility of accident propagation between units;
— IEs occurring in the site context and MU interactions (impact from one unit to another, e.g. due to radioactive release).

Certain simplifications made in SUPSA models might need revisiting to consider the MU context. For example, mechanical failure may be neglected in system analysis in SUPSA, whereas for shared systems these failures might not be negligible from a MU risk standpoint.

Thus, the preparation of Level 1 SUPSA models for the development of a MUPSA model includes the following steps, which are further elaborated in Sections 4.2.1–4.2.3:

(1) **Level 1 SUPSA model simplification:** This involves screening low risk contributors to the risk metrics previously selected;
(2) **Level 1 SUPSA model refinement:** This involves ensuring that the MU accident sequences are correctly modelled (e.g. modifying the PSA logic based on new IE grouping — see details in Section 4.3);
(3) **Level 1 SUPSA models development for all units:** This involves creating individual Level 1 PSA models for each unit to be considered in the MUPSA.

4.2.1. Model simplification

The methodology for the MUPSA is based on the careful application of screening to simplify and control the size and complexity of the model while still retaining the events and failures that are important to risk. This concept is well established in SUPSA, where screening is accomplished by considering the likelihood and impact of a given item (e.g. IEs, scenario, failures) relative to other items, and from that, determining which of those can be screened from the model without losing any important information about the risk profile of the unit being analysed.

For MUPSA purposes, the risk contribution for each hazard group in SUPSA in general can be used as an initial input to the screening process. Thus,

IAEA-TECDOC-1804 [3] addresses the screening out of MU events under special attribute HE-B06-S1, which states:

"For multiunit PSAs the screening of hazards meets one of the following criteria.
5) The individual hazards or correlated hazards do not have the potential to cause a multiunit initiating event.
6) An individual hazard or correlated hazards if subjected to detailed realistic analysis would not make a significant contribution to the selected multiunit PSA risk metrics, e.g. SCDF/SFDF, SLERF, SRCF, or SCCDF."[6]

Based on the above attributes, two criteria are developed to screen hazard groups from the MUPSA. These are discussed in the following subsections. Note that if Level 2 (or Level 3) PSA modelling is being performed, the screening considerations need to ensure the screening criteria are met for all levels of the analysis; for example, screened scenarios are demonstratively low risk for Level 1, Level 2 and Level 3 results.

4.2.1.1. Qualitative screening

Item 5 of HE-B06-S1 in Ref. [3] is consistent with Assumption 4 in Section 3. For this methodology, however, the impact is expanded to include the potential to cause a plant shutdown or degraded condition in another unit. The resulting qualitative criterion is as follows:

— **Screening Criterion 1 (qualitative screening):** A hazard group or IE can be screened out from the MUPSA model as a low risk contributor if all three of the following are true:
   (i)   The event does not immediately result in a trip of multiple units[7];

---

[6] SCCDF is the site complementary cumulative distribution function. This is the annual frequency of exceedance of consequences quantified in a Level 3 MUPSA for different consequence metrics such as early fatalities, latent cancer fatalities and property damage costs. Complementary cumulative distribution functions are aggregated to account for all the RCs and associated SU and MU accident sequences modelled in a Level 3 MUPSA. SFDF is the site fuel damage frequency; SLERF is the site LERF; and SRCF is the site release category frequency.

[7] When the MUPSA includes combinations of multiple POSs, since the plant could already be tripped, the steps involve consideration of whether the event causes a perturbation to steady-state operation that challenges plant control or safety. Additionally, for some hazards, a pre-emptive plant shutdown occurs in preparation for the hazard. Plant shutdown in response to a hazard is included in the MUPSA evaluation of plant trip.

(ii)   The event does not result in an immediate trip of one unit and a degraded condition at another unit that will eventually lead to a trip;

(iii)   The event does not result in a degraded condition at multiple units that will eventually lead to a trip of the units.

Applying the first screening criterion, it is not necessary to include IEs for the MUPSA unless the IE impacts multiple units; in other words, combinations of independent IEs on different units could typically be neglected (e.g. a loss of coolant accident (LOCA) at one unit simultaneous with a support system initiator at another unit).

When reviewing whether the IE impacts multiple units, if the MUPSA includes multiple POS combinations or locations (e.g. the SFP), the consideration of an IE would not just consider whether a plant trip occurs, but whether an IE occurs for the POS — such as causing a loss of decay heat removal.

4.2.1.2.   Quantitative screening

The quantitative screening criterion with regard to MUPSA is formulated as follows:

— **Screening Criterion 2 (quantitative screening):** A hazard group, correlated hazard, IE accident scenario or POS combination can be screened out from the MUPSA model as a low risk contributor if a bounding or demonstrably conservative estimate of CDF or fuel damage frequency (LERF) over the full range of hazard event severity is less than α% of the internal events CDF or fuel damage frequency (typical value of α is 0.1).

The quantitative screening criterion was applied for MUIE screening in the IAEA case study [11], and the ABWR MUPSA in the UK applied a similar screening criterion using the above Screening Criterion 2. In these studies, the IEs (both internal and external) could be screened out from the MUPSA if their contribution to the SUCDF was less than 0.1% of the total SUCDF. This is based on past MUPSA insights, as discussed in Ref. [1], that MUCDF is expected to be within one order of magnitude of the SUCDF, even when there is minimal sharing of SSCs among the units. Even if an event that contributed 0.1% of the SUCDF had a conditional MUCDF probability of 1.0 and all the higher CDF contributors had an average conditional MUCDF probability of 0.1, the most the screened event could contribute would be 1% of the total MUCDF, which is not significant. In both studies, the MUCDF was greater than 10% of the SUCDF and the use of the 0.1% value for α was supported by the above discussion. However, for other sites with a much smaller contribution from external hazards, a smaller value of α may be needed. In addition, the value of α needs to be verified in the MUPSA process, as described in Section 4.2.1.3.

If the scope of the MUPSA includes low power POSs, the screening criteria above can also be applied to the combinations of POSs. This may include consideration of the probability/frequency for the POS combination. This approach was successfully demonstrated in Ref. [38]. Meanwhile, screening needs to consider that shared or cross-tied SSCs used by multiple units could be unavailable as a result of maintenance during shutdown POSs.

Since each reactor unit could be in a different POS at the time of a MUIE, different combinations of POSs for each unit need to be considered. However, when considering many units, it is impractical to model all possible combinations of POSs [3].

An example approach was completed in KHNP's MUPSA project [37], in which a total of nine reactor units were considered. The details of this analysis, provided in Annex IV, introduced the concept of site operating states (SOSs) and suggested a screening approach in which five representative SOSs were selected based on historical experiences of overhaul and the long term plan for the overhaul schedules [47]. In addition, representative overhaul units and POSs were conservatively assumed on the basis of the conditional core damage probability (CCDP) to screen out certain POS combinations. In the case of a MUPSA, the SUPSA results can be treated as 'demonstratively conservative' (again, after ensuring the MU aspects are included in the SUPSA results) by assuming the CCDP or conditional large release probability for the other units on the site is 1.0 for all external hazards (examples of assumptions for internal hazards are presented in Ref. [52]).

Another option for simplification is screening the low risk accident sequences from modelled IEs, based on the SU model results. Screening low risk sequences from the original full scope SUPSA model can significantly reduce the size of the MUPSA model. In reviewing the SUPSA results, typically even risk significant IEs have only a few risk significant accident sequences (event tree sequences). The screening criteria need to consider that even if the second or additional units experience a high CCDP for those accident sequences, the screened out sequences would not become high risk contributors in the MUPSA.

For sites with more than two units or when the MUPSA includes modelling of multiple POS combinations, multiple screening criteria factors (α) may need to be developed. Additionally, the application of the screening criteria would need to consider which base CDF to use in the initial MUCDF screening level, considering that CDFs could be very different for non-identical units on the site.

Note that this screening only applies to the MU model used in the MUPSA. The SUPSA models, which are used to quantify SU core damage sequences, need to retain all their aspects without change.

4.2.1.3.   Verification of quantitative screening criteria

Once the MUPSA is complete, the validity of this screening criterion needs to be confirmed to ensure the screening criteria value used for α does not screen out potentially significant contributors.

Initially, this would involve a review of the final risk metrics (e.g. MUCDF or multi-unit large release frequency (MULRF)) against the quantitative screening criteria. Since the value of α is based on an 'expected' value for selected risk metrics (e.g. MUCDF/MULRF), if the calculated risk metrics is much lower than expected, the screening criteria α may need to be adjusted. It might need additional detailed analysis to verify whether previously screened out scenarios rise above the revised screening criteria. As a result, units with significant shared support systems, such as cooling water, electrical power or even containment, may need to include additional scenarios that could be initially screened out when applying the screening criteria set above.

Additionally, site specific aspects affecting the MUPSA need to be reviewed to ensure the screening criteria are reasonable. This may include any risk significant CCFs calculated in the unscreened MUPSA scenarios and whether these CCFs are applicable to scenarios just below the screening level. These CCFs would include the conditional probability of a conditional MU LOOP event following a trip. If the consequential LOOP is calculated to be high, some SU scenarios can transition into a MU scenario when the LOOP occurs. If the consequential LOOP is calculated to be above the value for α, higher risk SUIE screening using screening criterion 1 above may need to be analysed in detail.

Finally, if the MU risk only has significant risk contributions from a single hazard and the applied screening criteria result in most other potential MU scenarios being screened out from the MUPSA, lowering the screening criteria further needs to be considered. By lowering the screening criteria, additional risk insights may be gained from additional quantitative modelling. This would also ensure that if the significant risk contributing to the MU hazard was refined where the contribution was reduced, the MUPSA would not have to be refined further as a result.

4.2.2.   Model refinement

In this step, the SUPSA model is refined as needed to ensure the MU accident sequences are correctly modelled (e.g. modifying the PSA logic based on new IE grouping — see Section 4.3). The extent of the refinement performed depends on whether unique PSA models are available for all units on the site.

As such, some steps listed below might not be needed, as they were already performed when the individual PSA models were developed:

(1) Ensure the representative SUPSA model correctly accounts for MU aspects, as discussed in Section 4.3;
(2) Remove all random IEs that have been screened out or are not within the scope of the MUPSA (see Section 4.3.1), or remove all PSA logic that does not support the risk metrics selected for the MUPSA (see Section 4.1.2);
(3) Remove all internal or external hazard scenarios that do not have MU impacts (see Section 4.3);
(4) Remove all hazard induced IEs that are not correlated across the units (see Section 4.3);
(5) Regroup IEs as identified in Section 4.3.3;
(6) Modify the IE frequencies accounting for MU aspects (see Section 4.3.4);
(7) Modify the coding of the PSA model elements (e.g. IEs, BEs, fault trees) to reflect the unique number of units in preparation for creating a PSA model for each unit (if not already developed).

The process of creating these refined models is iterative throughout the remainder of the MUPSA tasks and is only completed once all of the tasks are completed.

The approach to creating a model for each unit is discussed in Section 4.2.3 below. Note that there can be slight differences in the refinement for the SU model depending on whether the single fault tree (SFT) method or master event tree (MET) method is used for the MUPSA. These two approaches are discussed in Section 4.4.2 and Appendix III, respectively.

Neither the IAEA case study nor the ABWR pilot MUPSA in the UK required any new system modelling or new BEs (other than HRA, CCF or fragilities). However, the MUPSA modelling could result in new or modified accident sequence analysis or modified system modelling, which would need to be performed prior to creating the MUPSA logic model.

Modified system modelling might include, for example, the expansion of undeveloped events crediting equipment on another unit. In addition, failure of the shared SSCs does not lead directly to core damage in the SUPSA, but it rather leads to MU core damage in the MUPSA. In this case, the inclusion of the SSC in the model might be reconsidered if the basis for screening was item 2 above. Modified accident sequence analysis could result from modified success criteria due to shared resources being utilized by both units affected by a MUIE. Any new or modified system modelling performed to support the above steps would be performed similarly to the original SUPSA. However, the BE probabilities or system modelling logic may need to be adjusted based on the MU impacts.

4.2.3. Development of models for all units

The MUPSA includes a combined logic model for all units on-site. The following provides some detail in creating the PSA models for all units on the site, if not already developed.

The IAEA case study provided some details on the steps needed to duplicate the PSA models for all units. The steps needed depend on what type of model will be quantified — MET or SFT — which is discussed in Section 4.4.2. The following steps are used to create the fault tree and event tree models use in either the SFT or MET approach:

(a) Duplicate the unscreened scope of the base PSA model(s) to create a PSA model (fault trees, event trees and data) for each similar unit, as applicable:
   (i) For sites with multiple groups of similar units, each base model for the similar units is used to develop the unit specific PSA model;
   (ii) If the accident sequence modelling is simplified (e.g. not all combinations are modelled), it may be possible that not all of the base PSA model will need to be duplicated.
(b) Recode PSA model elements to reflect the unique number of the unit, considering:
   (i) BEs or IEs with full dependency (e.g. the same event);
   (ii) Shared system and component BEs;
   (iii) MUIE naming.
(c) Merge the PSA models for each unit, making the proper interconnection of all shared systems, CCFs and seismic fragilities in accordance with the model integration steps discussed in Section 4.5.1.

When creating a MUPSA model for plants with shared systems, the model development needs to include a review of both the initial SUPSA model screening of SSCs and any simplified modelling such as modular or undeveloped events. Any SSCs screened out in the SUPSA that support multiple units may need to be added to the SUPSA models for each base model, depending on the potential risk impact of the SSC. Undeveloped events involving SSCs credited in multiple PSA models need to be expanded to ensure the combined MUPSA model correctly accounts for dependencies.

When creating a PSA model for additional units (if not available already), the applicability of the data (e.g. IE frequencies, failure rates, unavailability data) from the original unit PSA needs to be reviewed for applicability. For example, if the SUPSA includes plant specific data for the SU only (not site data), these data may not be applicable to other units. Similarly, for external hazard PSA models, if the fragility analysis was for a specific component for the original unit, this fragility may not be applicable to other units. When the data are found to be not applicable to other units, a new data analysis is necessary for these identified gaps.

When duplicating the PSA models for additional units, the portions of the model supporting screened out scope do not have to be duplicated to support the MUPSA model (see Sections 4.2.1. and 4.4.1). However, if the unit specific model will be used to develop SUPSA risk estimates for each unit, the model supporting the screened out scope needs to be duplicated.

When analysing MU events involving different (unscreened) POS combinations, the model development would involve unique PSA models for each unit. Existing PSA models often use different BE names for full power and shutdown, if the mission time is different. This would need to be addressed for shared systems, through either detailed modelling or a simplified approach. One possible approach would be to utilize the BE name for the longest duration mission time and use a multiplier (smaller than 1) for BEs with shorter mission times.

Shared systems between units will need to have consistent naming to ensure the dependency is correctly modelled in the MUPSA. Further discussion on modelling of shared systems is provided in Section 4.4.3. Shared components will have the same name for component related BEs that are the same between units. In some cases, the gate logic would be the same or shared between unit PSA models. In the case of cross-tie logic, the BE naming would need to be carefully reviewed to ensure that it correctly captured the dependency. For example, if backup of a system can occur by cross-tying the other unit's system, then the Unit 1 system logic would call the Unit 2 logic — and vice versa. When creating the Unit 2 model, this cross-tie logic would be reversed. However, the logic for both the Unit 1 and Unit 2 models would use the same BE names for each credited component, regardless of which unit the component was credited to supply.

Finally, the BE database supporting the PSA will eventually be merged to create a single database for the MUPSA. The above modifications would need to be performed and completed using a unit specific database. A comparison of each unit specific database between each model needs to be performed to identify issues created during model duplication.

4.3. IE ANALYSIS FOR MULTI-UNIT CONTEXT

The screening process of SUIEs described in Section 4.2 is a prerequisite for the analysis of MUIEs. The scope of MUPSA could include various types of MUIEs not screened out in the previous step of analysis. A detailed description with examples of types of MUIEs is provided below.

The task of MUIE analysis includes the following aspects:

— Review operating experience for gaps or changes in the SUPSA application to the MUPSA;
— Regroup any previously grouped IEs where the impact for SUPSA grouped IEs is different for MU, if this grouping will affect the scope of the detailed MUPSA;
— Assess the IE frequency for each identified IE and associated uncertainties;
— Document the MUPSA IE analysis.

For the purposes of this safety report, a MUIE is defined as an IE that immediately results in a trip or challenge to normal operation, or a degraded condition that eventually leads to a trip or challenge to normal operation, of two or more units. For the MUPSA, a degraded condition is one that affects either the systems required to prevent a trip of units or the systems credited in the PSA for mitigation of IEs, other than credited equipment in another unit.

4.3.1. Identification of various types of MUIEs

Each IE analysed in the SUPSA that is within the scope of the MUPSA is analysed for potential MU impact. According to the above definition of a MUIE, the MU impact considers the possibility of causing an IE on multiple units or a certain level of degraded condition[8].

If the MUPSA includes contributions from combinations of POSs or sources (such as the SFP), the identification of MUIEs needs to also consider how the POS impacts the definition of the IE and how the event impacts the reactor(s) or source, including degradation of equipment supporting the POS or source.

The factors affecting the IE identification for various internal and external hazard groups are discussed in the following sections.

4.3.1.1. Internal events

In the specific context of the MUPSA methodology, internal events can be classified into the following groups:

— LOCAs inside containment;
— LOCAs bypassing the containment (i.e. interfacing system LOCAs);

---

[8] In this context, the degraded condition implies that either systems required to prevent trip of units or systems credited in the PSA for mitigation of IEs are affected (which does not include administrative shutdowns — see Assumption 2 in Section 3).

— General plant transients;
— LOOP;
— Support system initiators.

The following are general considerations on MU impact for various categories of internal events:

(a) **LOCAs inside containment:** Typically, a LOCA in one unit could not result in a trip or degradation of multiple units. A review of the LOCA impacts needs to be done to confirm the potential impact of a LOCA on an adjacent unit.

(b) **Interfacing system LOCAs:** Typically, interfacing system LOCAs could not lead to a requirement to trip multiple units. However, these IEs need to be considered for inclusion in the internal flood hazard or other internal hazards (due to high energy line break effects) if the released liquid or steam could affect another unit. The likelihood of the propagation of the interfacing system LOCA impacting an adjacent unit would be considered as part of screening analysis.

(c) **Transients:** In most cases, transients (such as turbine trip, reactor trip or loss of feedwater) at any given unit will only impact each unit independently and will not result in a trip at another unit. However, this needs to be confirmed considering secondary effects of the IEs (such as submersion, temperature, pressure, spray, steam, pipe whip or jet impingement), which could potentially affect adjacent units through spatial interactions (e.g. in the case of a common turbine building).

(d) **LOOP:** From a MUPSA standpoint, four types of LOOP can be defined: plant centred, switchyard centred, grid related and weather related. In SUPSAs, these are often combined into a single IE, whereas in the MU context, separate representation may need to be used because the different LOOP types could lead to different MUPSA contributions (e.g. if grouped then regrouping is necessary, as discussed below). It is therefore necessary to consider the first two separately, whereas the last two may be considered together.

    (i) *Plant centred LOOP:* This involves LOOP on a SU and therefore does not cause an IE on multiple units, so it is not considered in MUPSA.

    (ii) *Switchyard centred LOOP:* This relates to LOOP caused by failures in switchyard. Inclusion of switchyard centred LOOP in MUPSA depends on the design of the site. If a given unit has its own switchyard (i.e. no interconnections with the switchyard of another unit until connection to the grid), then switchyard centred LOOP would only involve a SU and is not considered in MUPSA. In the case of switchyards being shared between units, switchyard centred LOOP needs to be considered for the shared units only[9].

(iii) *Grid related and weather related LOOP:* These events involve LOOP on one or more units on-site due to failures in the electrical grid or induced by extreme weather conditions. For these LOOPs, all units on the site could be affected simultaneously and therefore they need to be considered in MUPSA. Thus, MU sites might need modelling of multiple LOOP IEs in MUPSA. In addition, MUPSA might need to consider the potential contribution from consequential LOOP. In general, consequential LOOP is a relatively low probability in the SUPSA. However, for sites where the probability of consequential LOOP is relatively high or the BE for consequential LOOP is risk significant in the SUPSA, accident sequences involving a transient on one unit followed by a consequential LOOP may need to be considered in the MUPSA.

(e) **Support system initiators:** The impact of support system initiators is typically site and configuration specific and depends on whether the systems are shared or partially shared between units. These need to be considered in MUPSA if MU impact is possible.

4.3.1.2. Internal hazards

Consideration of internal hazards in MUPSA needs to include hazards that are typically considered for SUPSA, including the following hazards:

— Internal fires;
— Internal floods;
— Internal explosions;
— High energy line breaks;
— Turbine missiles;
— Drop of heavy loads.

The consideration of internal hazards starts with the determination of IEs that can cause MUIEs, as defined above. This includes scenarios with spatial interactions and scenarios of hazards affecting SSCs shared among units. In this

---

[9] For example, for a four-unit site, when each pair of units shares one switchyard that is independent of the switchyard for the other pair, the MUPSA model needs to consider the MU risk contribution from a switchyard centred LOOP for each pair but does not need to consider a switchyard centred LOOP at all four. Finally, if a single integrated switchyard exists for all four units, it is necessary to consider a switchyard centred LOOP for all four units.

context, units with shared areas or shared systems are of particular interest for the MUPSA. Most of the internal hazards in these areas would directly affect each unit. Additionally, areas in one unit that are adjacent to major sources of internal hazards (e.g. missiles or explosions) in another unit need to be reviewed. The impact of internal hazards could be essentially asymmetric, leading to different IEs at the affected units (i.e. it is possible that an explosion could cause a loss of normal alternating current (AC) at one unit and a reactor trip at the other). Therefore, for any retained internal hazard scenario, the grouping of scenarios for accident sequence analysis purposes needs to consider each of these combinations as a MUIE.

Some specific considerations for fires and floods are provided below:

(a) **Internal fires:** MUIEs induced by fires typically include scenarios involving multicompartment fires or scenarios of fires affecting SSCs shared among units. The MU fire analysis could require the implementation of new multicompartment fire analysis that was not performed within the SUPSA and/or an extension of the scope of existing fire analyses. However, the risk contribution of the SU fire scenario, as well as the probability of fire propagation, needs to be considered prior to expanding the SUPSA scenarios for MU contributions.

(b) **Internal floods:** MUIEs induced by floods typically include floods affecting common areas among units. When considering flooding IEs, it is important to avoid double counting of the events already considered in internal event MUPSA (e.g. steam line breaks in common turbine hall areas).

4.3.1.3. External hazards

In accordance with SSG-3 [49], the following external hazards cannot be screened out as an entire hazard category: seismic hazards, high winds and external floods. They therefore also need to be considered in a MUPSA. In general, the nature of external hazards means that in most cases they can affect more than one unit on-site. This includes scenarios with spatial interactions and scenarios of hazards affecting SSCs shared among units.

Treating external hazards in a MUPSA could be quite complicated, considering that they can cause multiple IEs in each unit on-site (e.g. LOOP, seismically induced LOCA and loss of support systems). In addition to functional dependencies between units, external hazards might also add complications related to potential hazard correlations across the units (see discussion on seismic correlation in Section 4.4.5.3). The basis for considering whether external hazard induced IEs need to be considered simultaneously for several units in the model is whether they are correlated. The idea behind this is not that the combinations of different IEs are not possible, but rather that they would be less significant to the total MUCDF.

Some specific considerations for seismic hazards, high winds and external floods are provided below:

(a) **Seismic hazards:** As mentioned above, the correlated IEs in multiple units will have a single fragility (e.g. large LOCA) and need to be considered for all correlated units on-site. Thus, a combination of different seismically induced IEs across multiple units would have independent fragilities and would be less likely (large LOCA and small LOCA). In addition to these IEs, seismic MUPSA need to consider IEs related to seismically induced loss of a shared system or structure.

(b) **High winds:** Consideration of high winds in MUPSA can be complicated, due in large part to the nature of the impact on the units. The analysis of over-pressure impacts is relatively straightforward. However, the impact of missiles generated by high winds can become complex because of the random nature of the impacts of missiles on SSCs. Thus, for scenarios retained for detailed analysis in the MUPSA (see Section 4.2), the MU analysis for high winds may need to be performed separately, but in a similar manner as the SUPSA — with new target sets involving all units considered in the analysis.

(c) **External floods:** Similar to the case of internal flooding, it is possible in the case of flood that there could be different IEs triggered at the affected units (i.e. it is possible that the flood could cause a loss of normal AC at one unit and a reactor trip at the other). Therefore, for any retained external flooding scenario, the grouping of scenarios for accident sequence analysis purposes needs to consider each of these combinations as a MUIE.

In general, many of the hazards listed in annex 1 of SSG-3 [49] could be applicable for inclusion in the MU model if they were included in the SUPSA model. The same considerations apply also to the combination of external hazards (e.g. aircraft crash and consequential fire or explosion, independently occurring external flooding and lighting).

4.3.2. Review of operating experience

The purpose of this task is to review operating experience data for the plant being studied, or similar plants, to determine gaps in the SUPSA that could miss key aspects of MU risk when applied to the MUPSA. Additionally, the operating experience review needs to identify important aspects that could influence the MUPSA, such as IEs, interunit CCF, units with shared design, operational weakness and human factors.

Typically, the operating experience review is available from the SUPSA study, including the collection of plant experience and history for applicable events (either from the plant being analysed or similar units). IAEA-TECDOC-1804 [3] lists multiple general attributes for which operating experience is necessary, including under POS analysis, hazard event analysis, IE analysis, system analysis, HRA, data analysis and dependent failure analysis. The results of the operating experience review need to be used to supplement MUIE analysis, and in general to determine any model changes or enhancements needed to apply the SUPSA models to the MUPSA model. It is expected that this review will also help to determine whether the proposed technical approach captures MU events of interest that have occurred.

4.3.3. Grouping of MUIEs

As discussed above, the grouping of IEs coming from SUPSA needs to be revised if they have different MU impacts. The following events are some typical examples:

(a) **LOOP:** If the SUPSA includes only a single IE and a single combined recovery curve, then the IE may need to be regrouped to include plant centred, switchyard centred, grid related and weather related LOOP events.
(b) **Interfacing system LOCA:** In the case that a subset of interfacing system LOCA events can impact adjacent units, they may need to be regrouped for MUPSA purposes.
(c) **Main steam line break:** The SUPSA may include the main steam line break in the containment, steam tunnel and turbine building. In such cases, typically only the break in the turbine building can potentially affect multiple units (e.g. in the case of a shared turbine building). Thus, the existing grouping of main steam line breaks might need to be revised for MUPSA purposes.

The concept of IE grouping is well known and understood for a SUPSA. A given group represents a set of IEs that have similar effects on the unit, similar demands on the unit systems, etc. Thus, IAEA-TECDOC-1804 [3] includes several attributes that impact the grouping of MUIEs, in particular the following ones:

> *"IE-C05: Each single unit common cause initiator and each multi-unit common cause initiator are considered as a separate group."*

"IE-C06: 'Multiple units initiators' affecting systems/equipment shared among several units are considered as separate groups."

Extending that principle to the MUPSA, IEs can be grouped by the same considerations plus the number of units affected and the IEs that occur at each unit. Therefore, for instance, two-unit LOOP events could be represented by one IE group and four-unit LOOP events by another one, etc. As with SUIE groups, each MUIE group would be represented by its own accident sequence model (see detailed discussion in Section 4.4).

4.3.4. Frequency assessment of MUIEs

The purpose of this task is a frequency assessment for all MUIEs to be considered in MUPSA. This includes estimating the frequency of any MUIE using and revising the frequency of a SU IE to recharacterize it as an IE that only impacts a SU. The IE frequency from the SUPSA can account for IE causes that impact both single and multiple units. As discussed in Ref. [1], the units for the MUIE frequency need to be expressed as a site-year.

As mentioned above, the MUIEs may involve the development of multiple IEs for the same type of event or hazard group (e.g. two switchyard centred LOOP IEs for a site — see Section 4.3.1.1). The frequency for each IE needs to reflect the frequency of the IE that has a MU impact[10], including consideration of severity or propagation, as applicable (e.g. two scenarios developed for a two-unit site to address a multicompartment fire — one starting in Unit 1 and the other starting in Unit 2). Furthermore, the total IE frequency needs to be retained when normalizing the IE frequency. For example, when separating LOOP IEs into MUIEs and SUIEs, the total frequency needs to be equal to the sum of the LOOP for each unit on-site.

Various examples of LOOP and seismically induced LOCA MUIE frequency assessment are presented in Ref. [1] for a two-unit site. Other examples of MUIE frequency estimation can be found in Korean studies [31, 37, 53], where all reactor trips that occurred in Korean NPPs over a 40 year period were reanalysed from the MUPSA perspective and each IE group (e.g. LOOP, loss of condenser vacuum, general transient) was classified into three categories: single, dual and all unit IEs. Frequency distributions on a per site-year basis were

---

[10] If the SUIEs are grouped, the IE frequency for the portion of the IE that has a MU impact will need to be evaluated. In addition, the frequency for the SUIE needs to be revised to reflect the part of the IE that impacts one and only one unit to permit integration of the results of the SUPSA and MUPSAs.

obtained from a Bayesian update of the Jeffreys non-informative prior with the obtained data (see Annex IV for more details).

The IE treatment in the MUPSA, including any simplification in the modelling, needs to be identified in the MUPSA documentation. This could involve adjustments in either the IE frequency or the final quantification, which can depend on the methodology used in the accident sequence analysis and quantification.

4.4. LEVEL 1 MUPSA LOGIC MODEL DEVELOPMENT

The purpose of this step is to develop the MU logic model for all risk metrics identified in Section 4.1.2 above. This step includes the modifications to the SUPSA logic models, including changes to the BEs, CCF modification, system logic model changes and changes to the HRA to account for MU accident sequences.

Once the individual PSA models are created for each unit on-site, the models are combined to create the MUPSA model. The approach used may involve any approach described in Section 4.4.2. The advantages and disadvantages of each are discussed below.

The development of the MUPSA model is based on the risk metrics and unit permutations/combinations within the scope of the MUPSA. Screening of permutations/combinations is possible in order to reduce the level of effort of the MUPSA, as described in Section 4.4.1. The changes to the system models, CCF, HRA, external hazard correlations or other data are based on the safety functions supporting the accident sequence logic created for the MUPSA model, as described in Sections 4.4.3 to 4.4.5.

4.4.1. Screening of MU combinations

In the context of accident sequence analysis for MUPSA, some possible combinations of unit core damage results could be screened out based on Screening Criterion 2, discussed in Section 4.2.1. The screening criteria for individual IEs can be used to support the screening of MU combinations. For example, if the dominant SFP PSA contributors and internal events PSA are not shared (e.g. no common MUIEs), then the MUPSA may involve MU SFP fuel damage, but the combination of SFP on one unit and internal events affecting the fuel in the vessel on the other unit can be screened out.

The consideration of combinations can be complicated for sites with more than two units or with multiple POSs within the scope of the MUPSA. For this step, all combinations of IEs within the scope of the MUPSA need to be reviewed, resulting in a determination of what combinations are within the scope of the MUPSA development discussed below. The integration of the individual SUPSAs, therefore, is impacted by both the unscreened MU scenarios and the unscreened combinations of these MU scenarios. Any screened out combination needs to be documented, including the basis for screening.

4.4.2. Accident sequence analysis

Once the scope of the MUPSA has been determined, including the permutations of MU scenarios, the MUPSA combined model is developed using the SUPSA models from each unit. The MUPSA integrated model can be developed using several techniques, building on the approach used for the SUPSA.

One approach used in previous MUPSA studies [1] is to develop combined event trees for each analysed MUIE representing all possible outcomes. For example, for a two-unit LOOP, the Unit 1 and Unit 2 LOOP event trees are combined to defined four possible outcomes: OK state, core damage on Unit 1, core damage on Unit 2 and core damage on both Units 1 and 2. However, although this approach is possible, for a site with more than two units, the size of the event trees can become quite large and difficult to manage. Given that, this approach does not fully meet the principles discussed in Section 3 (applicable to a site with many units).

Multiple approaches for MU accident sequence combination were identified in Phase I of the MUPSA project [54]. MET, SFT and hybrid approaches are discussed below. Other approaches not discussed below can also be used for MU accident sequence analysis (see the annexes for examples and approaches used in Member States).

(a) **MET approach:** The objective of the MET approach is to model the plant response (technological equipment, operator actions, etc.) for each IE in a single combined event tree for all units on the site. Appendix III discusses the advantages and disadvantages of the MET approach. As discussed above, the MET approach can provide results and insights for all possible combinations of unit risk/CDF. On the other hand, the MET approach needs each potential MUIE to be modelled and analysed separately, which can complicate the MUPSA analysis.

(b) **SFT approach:** The SFT approach, also sometimes referred to as a top logic modelling approach, combines risk logic for all units, all hazards and all modes under a single-top logic gate. If multiple risk measures are desired, then each risk measure can be modelled using an SFT top gate. As with the

MET approach discussed above, the SFT approach can provide results and insights for all possible combinations of unit risk/CDF.

(c) **Hybrid approach:** The hybrid approach involves a combination of MET and SFT approaches. First, a MET is developed with a single event tree node for each of the modelled units or sources. The node for each unit or source is then modelled with a unit specific top event or SFT. This methodology is described in Appendix III, Section III.3.

Appendix III provides more detailed information about each of these approaches and discusses the advantages and disadvantages of them.

4.4.3. System analysis

The purpose of MU system analysis is to model the MU impact on system analysis for each unit's PSA by modifying the original system models. In MUPSAs, the modelling needs to ensure the BEs used in each SUPSA model are consistent to ensure proper accounting for dependencies.

Although the availability of a shared SSC to each unit dynamically changes depending on the state of each unit, these dynamics are difficult to model using traditional static PSA methods. Moreover, it is often the case that the unit priority for the shared SSC is not described in associated operating procedures. Therefore, this approach designates the priority of using a shared SSC between or among units, considering the relative risks of the units, and modifies the fault trees related to the SSC according to the order of priority. Figure 3 shows example fault trees for an alternative AC diesel generator (AACDG), which is connected to a unit in case of station blackout (SBO) and is shared among four reactor units. Here, it was assumed that the priority for the AACDG was given in the order of Unit 1 → Unit 2 → Unit 3 → Unit 4. That is, in cases where SBO events simultaneously occur in two or more units, the AACDG is available to only the unit with the highest priority (e.g. Unit 1), while it is not available to the other units with lower priorities (e.g. Units 2, 3 and 4). Switching its connection from one unit to another was not credited. Korean studies provide an approach to modelling shared SSCs in MUPSA [31, 37, 53, 55].

Modelling of shared systems for most MUPSAs for a two-unit site could be simpler than the example shown in Fig. 3. In some cases, this may be the development of a common fault tree which supports both units' PSA models.

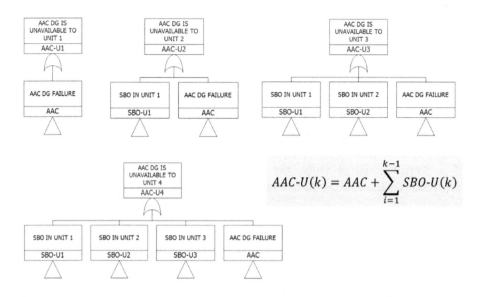

FIG. 3. Example fault tree model for an SSC shared between four units.

$$AAC\text{-}U(k) = AAC + \sum_{i=1}^{k-1} SBO\text{-}U(k)$$

4.4.4. HRA

The purpose of the MUPSA HRA is to assess and quantify the HEPs for MU accidents, including evaluation of the dependency of HEPs between operator actions being performed on different units. As with other steps in the MUPSA methodology, it is assumed the SUPSA HRA supporting a MU site includes MU considerations, such as the operator actions on shared systems or the impact of shared resources such as a shared control room. In the MUPSA HRA, the HEPs are assessed based on MU core damage and potential release events occurring, including severe accident conditions, when applicable.

The forthcoming IAEA Safety Report on HRA [14] discusses the MU considerations for HRA. This Safety Report discusses in more detail both the positive and negative aspects that another unit may have on the human actions performed during a MU event. This includes discussion on resources and interactions between units. The discussion below supplements Ref. [14] in providing details that can be used to quantify the HEPs used in a MUPSA.

MU accidents could create a specific context for human interactions, which needs to be considered in HRA when analysing HFEs for MUPSA purposes. The following list summarizes the specific factors for the MU context:

— Shared human resources between units;
— Shared control rooms (if applicable for a given NPP design);

— Increased stress due to the MU accident conditions;
— Impact on accessibility due to the degraded condition of other unit(s) on-site;
— Factors connected with severe accident conditions on other units at the site (e.g. radiological release, hydrogen detonation).

The overall HRA approach and tasks are similar to SUPSA, which includes the following steps as described in Ref. [14]: identification and definition of HFEs; qualitative assessment; quantitative assessment; and incorporation into the MUPSA model. HRA in MUPSA involves a phased approach with initial screening HRA (which estimates the HEPs conservatively), followed by detailed analysis. The steps for the HRA are discussed in the following paragraphs.

Identification and definition of HFEs. HFEs with potential MU impacts discussed above are identified. As mentioned in the general Assumption 14 in Section 3, preinitiator HFEs identified for systems not common to multiple units are assumed to be independent. As such, post-IEs are reviewed to determine which could be affected by MU impacts. For plants with significant shared and/or common areas, this assumption may need to be revisited.

Qualitative assessment. The SUPSA includes the evaluation of significant aspects of each HEP, including the following examples [49]:

(a) The timing of the action;
(b) The relevant plant procedures;
(c) The indications available to take appropriate action;
(d) The environment in which the action is carried out;
(e) Operational practices, for example, the structure of the operating crew and their responsibilities;
(f) The effect of prior actions on the action at hand;
(g) Information available to the operators, training they have undergone, etc.

For actions identified to have potential MU impacts, the above information is taken from the SUPSA for further quantitative evaluation in the steps below. In some cases, the base HEPs may remain unchanged from the SUPSA, but the dependency between units needs to be calculated, using a similar approach as used in the SUPSA HRA, based on similar procedures, training, etc. HEPs with longer timing may be impacted by performance shaping factors discussed above, including increased stress or radiological release, which may impact the SU HEP evaluation of the environment in which the action is carried out.

Quantitative assessment. Quantitative assessment includes both screening and detailed analysis of HFEs and consideration of potential dependencies. The purpose of screening MU HRA is to assign conservative HEPs for the

MUPSA, based on the MU contextual factors discussed above, without the need for detailed scenario specific HRA. Detailed analysis is foreseen for HFEs that are shown to be risk significant in the quantified MUPSA. In the stage of screening quantitative assessment, the allocation of shared personnel needs to be implemented with conservative assumptions, taking into account Assumptions 11 and 12 in Section 3. Additional consideration in the MU context needs to be made for the crediting of the technical support centre (TSC) for recovery actions. As the TSC has a site function, TSC credit needs to be applied as normal to all units.

One of the elements of a MUPSA is the concept that multiple units in (or progressing towards) a core damage scenario could impact operator actions at all units as a result of radioactive release from one of the units. This is generally a concern for MU scenarios where the timing is significantly different between units or for longer term actions where core damage on the first unit could have already occurred. Depending on the action and the release path, the local actions potentially affected by radioactive release may not fail but could become more difficult because of the need for protective gear or other precautions. In other cases, actions in the potential path of a very large release could have failed (e.g. HEP = 1) under release conditions on the first unit. The impact of radiological release needs to be considered in the potential increase in both stress and time to perform the action. The analysis needs to cover the potential impact of radiological release to the habitability of the MCR or local control stations credited in the MUPSA HRA. This evaluation includes consideration of any special protective gear for MCR operators following a release.

The case study presented in Appendix II did not show significant impact from radioactive release, due in part to most operator actions of importance likely occurring prior to core damage on any unit during a MU event. Additionally, many of the MU scenarios were dominated by MU CCF, and the impact of increased HEPs on the second unit was small. However, the observation shows that the impact of radioactive release from one unit to other units can potentially play a significant role depending on the risk profile. Thus, the impact of radioactive releases on HRA could be significant when the MUCDF results are not dominated by correlations between the units (e.g. in the case of different types of unit and/or in cases when the risk profile of one of the units is dominated by accident scenarios leading relatively quickly to core damage in comparison with other units).

Another challenge that is unique for the MU context is the consideration of interunit dependencies for identified HFEs. The dependency level is expected to be relatively high in the case of shared resources, when one or more of the operators supports the action or decision at multiple units. The dependency level is expected to be relatively low if the crews for each unit are completely self-contained (i.e. not sharing; an example of dependency analysis for such

HFEs is presented in the case study described in Appendix II). However, even in this context the potential dependencies connected with the TSC or another organization coordinating the activities on-site needs to be considered and could play a significant role in the results. Dependency analysis for internal and external hazards needs to consider any differences in impact of the hazard on each unit. The hazard can affect the units differently, especially if one unit has a higher hazard fragility. This can lead to different human performance between the units and result in less dependency between the unit specific HEPs.

In general, current HRA methods are applicable for implementation of HRA in MUPSA, given that MU contextual characteristics are properly analysed and reflected in HEP value quantification and in dependency analysis. Challenges and further efforts needed in this area are discussed in Section 6.4.

Incorporation of HRA in MUPSA model. Incorporation of HRA results in the MUPSA model is to be implemented using the same principles as for SUPSA. This typically includes incorporation of HFEs into event tree and fault tree models, modification of post-processing rules or logic models to account for dependencies and modification of the existing HEPs to account for the impact of the MU context.

4.4.5. Data, CCFs and fragility analysis

The purpose of the MUPSA data analysis is to calculate the modified or new BE probabilities used to calculate the MUPSA risk metrics, including the CCF probabilities and hazard fragilities.

As discussed in Section 4.2.2, if any new accident sequence or system analysis modelling is needed to support the MUPSA, this is performed using the existing SUPSA processing. The development of any supporting BEs, other than the MU HEPs, CCFs or fragility BEs, would be performed using the existing PSA processes.

The following sections provide discussion on data analysis, CCF and fragilities, and account for MU impacts.

4.4.5.1. Data analysis

In the case where the MUPSA includes new data analysis, the analysis is performed similarly to the analysis performed for the SUPSA. This may include, for example, data analysis associated with expanded modelling of shared systems that is modelled as a point estimate in the base SUPSA.

4.4.5.2. Interunit CCF analysis

In order to capture the potential for CCF in a MUPSA, it is necessary to extend the intraunit common cause approach used for SUPSA. The first part of that process is to determine at a high level when interunit common cause needs to be considered. This includes both the CCF IEs and the CCF affecting systems credited in the PSA. The CCF IE modelling is discussed in Section 4.3.1. CCF modelling for PSA credited systems is discussed further in this section.

The issue of MU CCF was first investigated in the two-unit Seabrook MUPSA [15], as summarized in Ref. [1], which includes the modelling assumptions and data analyses for emergency diesel generators (EDGs) and motor operated valves. To address this issue more broadly and to support the analysis of larger numbers of units, it is necessary to review and analyse potential CCF events with MU aspect data. More recent research has shown that a significant fraction of experienced CCF events have involved components in different units [44, 56].

Generally, the scope of the MUPSA CCF would be the same as the SUPSA CCF. That is, if the SUPSA does not consider that a set of components is susceptible to CCF within the unit, then it does not consider that it would be susceptible to CCF between units. However, there could be components where there is only one per unit (e.g. turbine driven auxiliary feedwater pump) where intraunit CCF would be included in the MUPSA.

An approach for the MUPSA is to initially model CCF in a conservative and simplified approach and to perform detailed CCF analysis for risk significant CCF events only. Transition from simplified to detailed CCF includes the following steps:

(1) Model interunit CCF for each component group using a simplified and conservative CCF BE. This single CCF BE is inserted into the system model for each affected unit in a location that affects all components within the CCF group. The SU CCF remains unchanged.

(2) For risk significant interunit CCF, perform detailed CCF modelling, accounting for component combinations for all components, in a similar manner to the SUPSA CCF modelling. The SUPSA CCF is replaced with this revised detailed CCF modelling.

(3) Interunit CCFs that remain risk significant after detailed modelling can be modified by reviewing the detailed failures in the CCF data to refine the CCF probabilities. The review needs to determine which CCF events would be applicable to a MU failure and which would not be applicable. New CCF factors are then developed, which results in the calculation of separate interunit and intraunit CCF BEs. These CCF BEs are then modelled in the MUPSA.

In developing the interunit CCF, the applicability of CCF needs to be considered. The MUPSA modelling includes several factors that can be unique to the MUPSA. These factors are discussed in the following paragraphs:

(a) **Different types of units:** If the units on the site are of different types (e.g. pressurized water reactors (PWRs) and boiling water reactors), then there is typically no technical basis for modelling CCF between the units. This is a simple extension of the basic CCF principle used in SUPSAs, whereby CCF is not considered for similar components in different systems. Furthermore, this approach could also be typically applied in the case of differences in reactor vendor, even for the same technology (e.g. CCF would not typically be considered between a Westinghouse PWR and a Framatome PWR located on the same site).

(b) **Different model components:** Current practice for SUPSA is that CCF is to be considered if the component has the same manufacturer and model number. Although there could be certain similarities in the overall design for different model components, it is likely that there will still be enough differences in the details (e.g. component function, specification, performance, manufacture, materials) that CCF is not to be considered.

(c) **Different vintage components:** Individual component technologies and manufacturing techniques change over time, and this will affect the similarity of the equipment even if the performance specifications and all other aspects appear the same. On this basis, it is expected that components of different vintages could be less susceptible to CCF. Perhaps more subtle are nominally identical components (e.g. make and model number) manufactured at significantly different times.

(d) **Passive equipment:** The root causes and coupling factors for CCF of passive equipment are quite different from those for active ones (e.g. failures such as CCF of heat exchangers due to plugging are not typically connected to the design and manufacture of the equipment). However, the specific couple factor for passive equipment CCF in raw water systems (which already needs to be considered in the SUPSA) is connected to draw water used from the same source. In this case, potential CCF due to corrosion, intergranular stress corrosion cracking and other failure modes needs to be considered. For advanced plants (e.g. non-LWRs), passive reliability can be influenced by factors common to multiple units, for example, for plants that use a reactor vessel auxiliary cooling system, the reliability of this system is impacted by the outside air temperature [57] (see more discussions on extreme environmental stresses below). Thus, when calculating MU CCF for passive systems, the impact of influence factors that may be common between units needs to be considered in determining whether the CCF is considered in the MUPSA (see more details in Section 6.3).

(e) **CCF due to extreme environmental stresses:** The underlying CCF data need to be reviewed to identify CCF causes resulting from extreme environment stresses that could impact components on different units if they share these conditions. For example, some EDG CCF events are due to extremely cold weather, which could impact EDGs with diverse design principles or vintages.

Experience gained from the IAEA MUPSA case study (see Appendix II) and other MUPSA studies worldwide has shown that CCF could play a significant role in the site level risk profile. Therefore, the assumptions presented above need to be applied with detailed explanation and documentation. The listed assumptions and expectations might not always be applicable. Therefore, CCF coupling factors need to be analysed carefully and the assumptions are to be verified through sensitivity analysis.

As discussed above, the initial modelling of the interunit CCF is based on a simplified approach, which involves being conservative. When applying a conservative CCF model for 'n-of-n' combinations, the CCF probability needs to be estimated by bounding all combinations of MU CCF. For example, in the two-unit, four-train example above, if a 4-of-4 CCF event is included, the assigned probability needs to include the 3-of-4 combinations for the initial conservative modelling. Additionally, if the original SU CCF (in this case, 2-of-2 CCF) is retained, then the MU CCF would not need to include contributions from the SU CCF combinations.

If the simplified CCF model results in significant conservatism, detailed modelling needs to be performed by modelling each combination for each significant component group. As with the simplified approach, the CCF modelling will need to consider the modelled combinations of units and MUIEs.

When considering many units in MUPSA, the simplified beta factor approach can estimate a given risk metric very conservatively. In particular, it is very likely that the most severe case of all units failing is significantly overestimated, with the intermediate cases of 2 to $(n-1)$ units failing subsequently underestimated [35]. On the other hand, detailed modelling of all possible CCF combinations (e.g. using the alpha factor model) can produce fault trees too large to be quantified in a reasonable time and make the estimation of the necessary CCF parameters more complicated [31]. For this reason, in KAERI's recent study [35], a pragmatic approach to modelling CCFs was proposed for application to MUPSA involving six or more NPP units. The approach is basically a hybrid between the alpha factor and beta factor models.

Regarding the selection of CCF combinations to be modelled, this approach employs separate strategies for intra- and interunit CCFs. For intraunit CCFs, it does not change the combinations included in the SUPSA model for each unit. For interunit CCFs, though, the approach models all possible 'unit combinations' at the unit level, and thus a single interunit CCF BE is modelled for each unit combination, regardless of the number of components in each unit. Figure 4 shows an example of applying this approach to a three-unit case where each unit has three trains of components [35].

As for CCF parameter estimation, the approach assumes that the interunit CCF events to be modelled are subsets of the intraunit complete CCF event in which all the components in a SU fail. Therefore, each interunit CCF parameter is calculated using its fraction, which is evaluated by considering two variables: a 'component specific parameter' reflecting its characteristics and 'interunit CCF correlation' between the units included in the combination. Here, interunit CCF correlation is determined for a specific component type by using a decision tree as shown in Fig. 5, which considers three types of CCF coupling factors (hardware, operational factors and environmental factors) [35]. In this tree, the correlation assigned to each case is based on the recently published International Common Cause Failure Data Exchange (ICDE) Project report [58], which analysed 87 'multiunit CCF' events from the ICDE database for a wide range of component types and showed that hardware related coupling factors accounted for 52% of these events, followed by operational factors (33%) and environmental factors (15%).

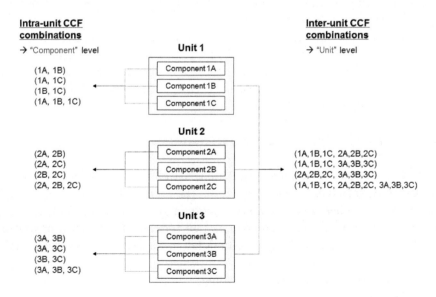

FIG. 4. Example CCF combinations to be modelled (three units, three trains per unit) [35].

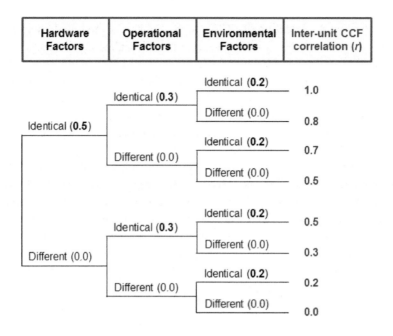

Hardware Factors	Operational Factors	Environmental Factors	Inter-unit CCF correlation (*r*)

Identical **(0.5)**
 Identical **(0.3)**
  Identical **(0.2)** — 1.0
  Different (0.0) — 0.8
 Different (0.0)
  Identical **(0.2)** — 0.7
  Different (0.0) — 0.5

Different (0.0)
 Identical **(0.3)**
  Identical **(0.2)** — 0.5
  Different (0.0) — 0.3
 Different (0.0)
  Identical **(0.2)** — 0.2
  Different (0.0) — 0.0

FIG. 5. Interunit CCF correlation decision tree [35].

By considering interunit CCF correlation, the proposed approach [35] makes it possible to deal with 'non-identical but partially correlated' components and their asymmetrical relationships.

The potential impacts of the MU CCF described above in this section need to be reviewed while reviewing the MUPSA results. Cut sets need to be reviewed to identify combinations of SU CCF BEs where the potential for MU CCF has not been considered. Sensitivity analyses need to be performed to note the impact of CCF modelling assumptions on the results.

4.4.5.3. Interunit external hazards fragility dependencies

The concept of intraunit correlation is well established in SU external hazard PSA, including seismic PSA [59]. The typical approach is to consider SSCs to be either fully correlated (correlation probability 1.0) or fully independent (correlation probability 0.0). Full correlation is assumed when a set of SSCs meets all the following conditions (all that apply to the external hazard):

— They are located in the same building;
— They are located on the same level;
— They are essentially identical;

— They are orientated in the same direction (for seismic PSA).

The intraunit correlation criterion is based on a simplified approach to considering whether the SSCs are subjected to approximately the same external hazard or seismic demand and have approximately the same external hazard or seismic capacity. The first two conditions speak to the demand side and the second two speak to the capacity side. In extending the concept to interunit correlation, it is necessary to add conditions that look at the special situation of structures associated with two different units.

There is really no difference between SU and MUPSA on evaluation of the capacity side. It is still an issue of whether the SSCs are essentially identical and are orientated in the same direction. This will only apply if the units are of the same type, design and vintage. It is unlikely that units that differ in these factors will have SSCs with close enough capacity to be considered correlated.

Generally, the correlation groups used for the SUPSA can be applied to the MUPSA directly, without additional grouping. However, where the SUPSA credits components containing only one component of a specific type (e.g. turbine driven auxiliary feedwater pump or reactor core isolation cooling pump), new MU correlation groups would need to be developed. It is expected that new correlation groups would be similar to any new CCF modelling discussed in Section 4.4.5.2 above.

Because the structures that house the equipment will be at different locations on the site, additional considerations may be needed to obtain a representative fragility for those cases.

For MUPSA, one approach for interunit fragility correlation analysis is similar to the CCF analysis, starting with a conservative approach followed by more detailed analysis. This approach includes the following steps:

(1) For identical components with the same seismic demand, assume the component fragilities are fully correlated across units; otherwise, assume that they are not correlated. This is typically similar to the approach for SUPSA. Other conditions discussed below, such as orientation or location on the site, are not initially considered.

(2) Upon completion of the initial MUPSA seismic and other external hazard analysis, perform a sensitivity analysis considering variations to the correlation. The IAEA case study, which included a MU seismic PSA, assumed zero correlation — which only reduced the MUCDF for seismic hazards (two-unit case) by ~30%. The ABWR seismic PSA pilot analysis in the UK showed a slightly higher reduction in CDF, but the analysis further identified that a large number of components would need to be subject to detailed correlation analysis to obtain a significant reduction in MUCDF.

(3)   Review the other factors discussed below for significant component groups to determine whether any correlations can be qualitatively screened out as non-correlated. This step is performed in a similar way to the SUPSA. For example, if the components on different units are located in different structures and orientated differently, the MUPSA can assume no correlation in the seismic PSA.

(4)   If the sensitivity analysis identifies that a significant reduction in MU risk can be obtained by analysing the correlation in more detail, this analysis is performed for a limited number of component groups. If the detailed correlation analysis shows that the highest risk significant component groups remain highly correlated, additional correlation analysis may not be needed. The amount of effort given to correlation refinement needs to be optimized to ensure the analysis effort is not wasted on non-risk significant changes to the results.

Upon completion of the above steps, the MUPSA external hazard and seismic PSA is quantified, with some additional sensitivity analysis for additional correlation analysis (e.g. to support that no additional refinement is needed). Simplified methods as described in Ref. [14] can be used (e.g. split fraction) in sensitivity analysis.

NUREG/CR-7237 [60] provides one approach to performing detailed seismic correlation analysis, which can be applied either to a SUPSA or a MUPSA. Ref. [60] can be used in support of Step 4 above. Performance of a detailed correlation analysis requires expertise in seismic fragility analysis and may require walkdowns (and detailed review of design documents for a plant in design) to ensure the factors affecting correlation are fully understood and documented. An example application of this methodology is provided in IAEA Safety Reports Series No. 96 [1].

4.4.5.4.   Detailed discussion on seismic correlation factors

For a MU seismic PSA, four types of situation requiring consideration of seismic dependencies between component failures are considered further in this publication:

(a)   **Different components at different units, with different seismic demand and different capacity:** The seismic responses at the mounting location (or floor) of these components will be different. Because the components are dissimilar, they will respond differently to the input floor motion, and their failure modes could also be different. These components are treated as fully independent.

(b)   **Different components at different units but the same seismic demand:**
The seismic responses at the mounting (or floor) of these components are
expected to be correlated because of a single earthquake ground motion
input to the buildings, and the same is expected in structure response spectra
at the location of the components. Because the components are dissimilar,
they will respond differently to the input floor motion and their failure
modes could also be different. Even with the same seismic demand as in
the first case, current quantification methods consider these components
as fully independent and are therefore judged to be generally appropriate.
However, if the components have approximately the same fundamental
frequencies and similar failure modes, a case might be made that partial
dependency exists.

(c)   **Identical components at different units and different seismic demand:**
The seismic responses at the mounting locations of these components may
be partially correlated because of a single input earthquake ground motion.
Because the components are identical, they will respond similarly to the
input floor motion and their failure modes could be similar. The variation
in responses can be large enough to minimize the impact of dependencies,
but this will vary considerably from case to case. Therefore, the decision
on treating these components as fully independent or dependent needs to be
taken on a case by case basis. The analyst will need to apply judgement to
ascertain whether using partial dependency is necessary.

(d)   **Identical components with the same seismic demand:** This is a common
situation requiring careful consideration of dependencies. It is possible to
have identical electrical or mechanical equipment across different units.
Also, it is possible for seismic demand for these components to be similar
or identical in the frequency range of interest. In these situations, treating
these components as fully dependent (i.e. 'one fails, all fail') appears to be
reasonable and not overly conservative.

The probability of a cutset with correlated seismically induced failures is
calculated by assessing the multivariate probability distribution for dependent
failures. In the case where fragilities and responses are log-normal variables,
the above mentioned distribution might be used to calculate the joint failure
probabilities (see Annex IV), as follows:

$$P_{12\ldots n}(a) = \int_{-\infty}^{\ln\left(\frac{a}{A_{1m}}\right)} \int_{-\infty}^{\ln\left(\frac{a}{A_{2m}}\right)} \ldots \int_{-\infty}^{\ln\left(\frac{a}{A_{nm}}\right)} \frac{1}{\sqrt{|C|(2\pi)^n}} \exp\left(-\frac{1}{2} x^t C^{-1} x\right) dx \qquad (1)$$

where

$P_{12\ldots n}(a) = P\left(\cap_{i=1}^{n} A_i < a\right)$ is the combined probability of seismic failures,

$x^t = \begin{bmatrix} x_1 & x_2 & \ldots & x_n \end{bmatrix}$ is the random variables vector,

$$C = \begin{bmatrix} \beta_1^2 & \beta_{12}^2 & \ldots & \beta_{1n}^2 \\ \beta_{21}^2 & \beta_2^2 & \ldots & \beta_{2n}^2 \\ \ldots & \ldots & \ldots & \ldots \\ \beta_{n1}^2 & \beta_{n2}^2 & \ldots & \beta_n^2 \end{bmatrix}, \beta_{ij}^2 = \mathrm{cov}\left(X_i, X_j\right), \text{ is the covariance matrix.}$$

To determine the covariance matrix in cases where data regarding response and capacity correlation are not available, assumptions regarding the correlation between seismic fragilities are needed.

Currently, industry practice considers the seismic correlation level among component failures in a simple way, because of (a) the difficulty of determining correlation groups and correlation levels and (b) the difficulty of calculating combination probabilities of correlated seismic failures. The system analysts assign extreme values of 0 and 1, for 'fully independent' and 'fully dependent', respectively, by using their judgement on correlation among seismic failures.

This approach contributes to the uncertainty in a CDF for seismic hazards. Depending on the combination of logical operators in fault trees, the seismic CDF could be either overestimated or underestimated. There has been a need to develop an approach to modelling seismic correlation with seismic CCFs explicitly. Such an approach would make it possible to avoid the above mentioned uncertainty.

The conversion of the correlated seismic failures into seismic CCFs is introduced in Annex IV. If the partial correlation among seismic failures is defined in the covariance matrix and the correlated seismic failures are successfully converted into seismic CCFs, the conservative MUCDF can be replaced with a more accurate MUCDF. The case study discussed in Appendix II evaluated the impact of partial correlations between units. The analysis necessary for reducing conservatism in correlations between units can be quite extensive. The case study showed that this level of effort may not substantially reduce the MUCDF.

4.5. LEVEL 1 MUPSA MODEL INTEGRATION AND QUANTIFICATION

The purpose of this step is to integrate the model developed and modified as described in Section 4.4 and to quantify the MUPSA model for risk metrics identified in Section 4.1.2.

4.5.1. Model integration roadmap

The model integration is performed to combine the MUPSA logic model, discussed in Section 4.4.2, with the revised system models discussed in Section 4.4.3. These PSA models are updated to include revised HRA, CCF, correlations and data, as described in Sections 4.4.4 and 4.4.5.

The MUPSA model integration steps depend on the MUPSA logic model method used (see Section 4.4.2) and the software used for the PSA. As described in Section 4.4, the logic models for internal and external hazards can be developed and quantified separately. If this is the case, the quantification of the risk metrics would require a separate risk aggregation step [5]. When the MUPSA includes multiple POSs, the model integration and quantification below involves combining the logic from the unscreened POS combinations.

A specific issue is the integration of the SUPSA and MUPSA results, which typically could be implemented after completion of the MUPSA quantification described in Section 4.5.2. If the PSA scope includes Level 3 study, then the integration in the Level 3 MUPSA model is the most accurate and comprehensive, since integration of the risk metrics related to off-site consequences is straightforward. However, if the PSA is limited to Level 1 or Level 2, then the integration involves adjustment to the SUPSA results, either by modifying and revising the SUPSA modelling and quantification or by mathematical adjustment of the SUPSA results (if needed, e.g. for importance measure analysis). The roadmap for SUPSA and MUPSA model integration includes the following steps:

Step 1. SUPSA model refinement.

Step 2. Modification of the SUPSA model to correctly account for MU aspects:

(a) Adjustment of IE modelling;
(b) Adjustment of MUPSA accident sequences;
(c) Adjustment of CCF modelling.

Step 3. Quantification of SUPSA for all relevant IEs (SUPSA risk profile).

Step 4. Quantification of MUPSA for all relevant results (MUPSA risk profile).

Step 5. Aggregation of SUPSA and MUPSA results (SUPSA+MUPSA risk profile).

The steps included in the roadmap for SUPSA and MUPSA model integration are briefly elaborated below.

Step 1: SUPSA model refinement

In general, the SUPSA model needs to be modelled in such a way as to allow for the results to be quantified as SU risk only (e.g. SUCDF) or as the total unit risk (e.g. unit CDF), depending on the analysis. This can be performed using any number of modelling techniques, including post-processing of the quantification or the use of NOT gates to remove MU contributions. Separate SU and MU accident sequence models can be maintained. If the MET approach or hybrid approaches are used (see Appendix III), the SU risk results may be a direct output from the MU accident sequence modelling.

Step 2: Modify the SUPSA model to correctly account for MU aspects

Refinement of the SUPSA model to include MU considerations is already discussed in detail in Section 4.2 and further elaborated in sections related to the specific PSA tasks. Some specific aspects related to the PSA tasks in regard to the integration of SUPSA and MUPSA results are presented below.

(a) **Adjustment of IE modelling.** As discussed in Section 4.3, the MUPSA modelling involves reanalysis of the IEs included in the base SUPSA, including potential regrouping of the IEs, reassessment of the IE frequencies and unique modelling for the IEs in the MUPSA. For SUPSA modelling, the revised grouping, frequencies and modelling need to be incorporated into the SUPSA for each unit. One complication involves the consideration of POS combinations in the MUPSA, which needs to be included in the SUPSA. This modelling would differ depending on whether the MUIE is modelled using a calculated frequency or if the IE is calculated on a yearly basis and adjusted with an availability factor (e.g. fraction of the time all units are at power).

(b) **Adjustment of MUPSA accident sequence modelling.** Section 4.3 provides an approach to modifying the SUPSA model, including accident sequence and fault tree modelling, in support of the MUPSA. Depending on the approach, this modelling can be used in support of the SUPSA. For example, if the accident sequence modelling for the MUPSA is performed using the MET approach discussed in Appendix II, the success branches can be used to model the SU-only CDF.

(c) **Adjustment of CCF modelling.** One important aspect is the modelling of dependencies. The revised CCFs need to be incorporated into the SUPSA when integrating SUPSA and MUPSA.

Step 3: Quantification of SUPSA for all relevant IEs

The objective of this task is the derivation of SUPSA results (risk metrics, e.g. SUCDF or SU LERF, and the relevant risk profile) by quantification of the SUPSA model. The process of quantification is well known and described in IAEA Safety Standards Series Nos SSG-3 [49] and SSG-4 [50] and in an ASME/ANS standard [16].

Step 4: Quantification of MUPSA for all relevant results

The objective of this task is the derivation of MUPSA results (risk metrics, e.g. MUCDF or MU LERF, and the relevant risk profile) by quantification of the MUPSA model. The process of MUPSA quantification is described below in Section 4.5.2.

Step 5: Aggregation of SUPSA and MUPSA results

The site risk metrics (e.g. site CDF) would include the combined CDF for each of the SUs added to the MUCDF results. If the site included three or more units, the site CDF would include contributions from each of the unique MUCDF combinations. This quantification can be used to develop unique importance measures, such as important SSCs for the site CDF or for unique unit combinations.

If the quantification does not need to include the calculation of importance measures, the calculation of risk metrics such as site CDF and SU-only CDF can be estimated by calculation without model modification. However, the numbers may not fully account for all aspects of the MU modelling, including the impact of dose release on the SUPSA results. A review of the contribution of dose release to the MUPSA results is suggested if this approach is used.

Finally, the aggregation of the MUPSA and SUPSA results is relatively straightforward in the Level 3 PSA, since the risk measures, including the dose assessment, can be easily integrated into a single result. However, prior to integration, the SUPSA results will need to be modified as discussed above to remove MUPSA contributors to the base unit specific PSA models. Once this is complete, all contributors to site risk or specific combinations of units (e.g. a two-unit combination for a site of three or more units) can be integrated. Importance measures can then be calculated using a similar approach to the Level 1 and Level 2 PSA.

4.5.2. Model quantification

The quantification process involves three steps:

(1) Quantification of the MUPSA for each hazard for the SU and MU risk metric;
(2) Aggregation of the quantified results for each hazard for the SU and MU risk metric;
(3) Aggregation of the MUPSA results with the SUPSA results, as needed, to quantify site risk metrics.

The quantification for each hazard is performed using a similar process to the existing SUPSA. The quantification may involve multiple solution steps, if multiple risk metrics are selected for the MUPSA. For example, for a three-unit site, the quantification could involve separate quantification for a two-unit CDF, as well as quantification of a three-unit CDF.

The quantification needs to account for the limitations in the initial modelling for MUIEs. If, for example, the MUPSA includes IEs initiated at one unit (but having a MU impact) in only one of the PSA models, the final quantification needs to account for the simplification for each risk metric.

Once the PSA model for each hazard has been quantified for each unscreened POS combination, the results are combined by providing a combined set of cut sets for each risk metric. The MUPSA quantification for each risk metric needs to account for all combinations of unit sequences, which can be missing or modelled as representative sequences in the MUPSA model. This modelling approach is discussed in Section 4.4.2. For example, if the MUPSA models a SUIE that could degrade another unit and the modelling includes only one of the two possible IEs, then the missing combination needs to be accounted for in the quantification. As noted below, this can result in a reasonable estimate for the risk metrics but may affect the importance measures. Another example would be whether the MUPSA modelling includes MU sequences addressing all possible ordering of core damage and releases.

The risk metrics identified in Section 4.1.2 can be quantified using the combined results. The quantification of risk importance measures can also be performed (discussed further in Section 4.6.2), although, as mentioned above, this quantification can be complicated depending on whether the results from all hazards can be combined and whether there are limitations or simplifications in the IE analysis.

The quantification for risk metrics for sites with three or more units may involve the modification of the results to account for specific unit combinations. For example, in the three-unit site example, when calculating the two-unit CDF, the results for the three-unit CDF may need to be removed — unless the two-unit CDF results are presented as two or more units.

Additionally, the quantification of the MUPSA risk metrics needs to consider the potential overlap and double counting of risk, such as the percentage of SUCDF results that include MUCDF (see Section 4.5.1 in Ref. [1]). For example, for a two-unit site, with high correlation for seismic events, it may be that a majority of the CDF contribution may involve a MU interaction. This is demonstrated in the IAEA case study, where more than 80% of the CDF risk for Units 1 and 2 involved a two-unit CDF (note that the case study conservatively assumed high dependence between units and was slightly simplified compared to a full seismic PSA). In this case, the frequency for a SUCDF event (only) due to a seismic event would be much lower than the frequency for a MUCDF.

Looking at the IAEA case study further, since the site involved four units, the quantification for selected risk metrics would need to consider the overlap between the four-unit CDF, each two-unit CDF combination and the SUCDF for each unit. This can become complex for even CDF calculations (without considering other metrics). This quantification can be addressed either in the initial quantification by removing higher level (number of units) contributions from the lower level combinations, or by later adjusting the frequencies mathematically. The latter method is an easier approach. However, this adjustment would complicate any risk importance measure calculations, since these are based on the quantified results for each combination. The IAEA case study did not fully quantify each risk metric and did not attempt to adjust the lower level CDF results by removing contributions from higher level (unit) combinations.

The quantification of risk metrics would involve the quantification of MU metrics (e.g. MUCDF), SU metrics (e.g. SUCDF) and site risk metrics (e.g. SCDF). As mentioned above, the calculation of each of these metrics has to account for overlap or double counting. As an example, the following provides the results for the IAEA case study for LOOP involving Units 1 and 2 (denoted U1 and U2 in the below) only (note that the case study included four units):

U1 CDF = 1.13×10^{-6}/year
U2 CDF = 1.13×10^{-6}/year
U1 and U2 CDF = 2×10^{-8}/year

In this case, the MUCDF is directly quantified at 2×10^{-8}, or less than 2% of the base Unit 1 and Unit 2 CDF. The SUCDF (SU only) is then calculated for each unit:

U1 (only) CDF = 1.13×10^{-6}/year $- 2 \times 10^{-8}$/year = 1.11×10^{-6}/year

U2 (only) CDF = 1.13×10^{-6}/year $- 2 \times 10^{-8}$/year = 1.11×10^{-6}/year

Finally, the site CDF is calculated by combining the MUCDF and SU (only) results:

SCDF = U1 (only) + U2 (only) + MUCDF
= 1.11×10^{-6}/year + 1.11×10^{-6}/year + 2×10^{-8}/year = 2.24×10^{-6}/year

Another challenge is the minimization of quantification uncertainty in MUPSA modelling. If SUPSA modelling and quantification methods such as negates and delete-term approximation (DTA) are directly employed for MUPSA, the MUPSA risks such as seismic MUCDF and exposure doses in Level 3 MUPSA can be overestimated or distorted. In Section IV–4 of Annex IV, examples for the modelling and quantification of uncertainty sources, as well as methods to minimize quantification uncertainties, are provided. The examples in that section represent the lessons learned from the Korea Foundation of Nuclear Safety project. Examples of quantification uncertainty sources and potential approaches to minimize them are summarized in Table 1 below.

4.6. ANALYSIS AND INTERPRETATION OF THE LEVEL 1 MUPSA RESULTS

Risk integration and interpretation of the MUPSA results, including the sensitivity analysis and uncertainty analysis, is performed in a similar manner as for the SUPSA.

TABLE 1. MUPSA QUANTIFICATION UNCERTAINTIES

MUPSA	Quantification uncertainty sources	Approaches to minimize quantification uncertainty
Seismic MUPSA	Multi-unit level negates	Avoiding modelling negate Application of post-processing (see Section IV–3.3 of Annex IV)
	Full or zero seismic failure correlation	Conversion of correlated seismic failures into seismic CCFs (see Section IV–4.3 of Annex IV)
Level 2/3 MUPSA	Assumption that all NPPs are located in a single position	Application of the multiple location method (see Section IV–3.4 of Annex IV)

As discussed in IAEA-TECDOC-1804 [3]:

"The objective of the results analysis and interpretation activity is to derive an understanding of those aspects of plant design and operation that have an impact on the risk. In addition, an important part of this task is to identify the key sources of uncertainty in the model and assess their impact on the results."

Quantification of the MU results needs to include the calculation of the MU/SU ratios for each selected risk metric. The ratios calculated in the IAEA case study are discussed in Appendix II. For example, for a two-unit site, the ratio of MUCDF to SUCDF would be calculated directly from the results. A high ratio (e.g. above 0.1) would indicate a high conditional core damage for a second unit if the first unit experiences a core damage event. This ratio can be calculated for each risk significant hazard or IE.

To perform interpretation of the results, the MUPSA quantification and a SU sensitivity analysis would investigate the following:

— Examine results for underestimation or overestimation of risk, including asymmetry between units.
— Interpret importance measures — use with caution because of model simplification, but can be used to check MU modelling balance (i.e. a component's importance on one unit needs to be comparable to the same component in the sister unit).
— Perform uncertainty analysis on documented assumptions (model simplification may preclude a viable parametric uncertainty analysis).
— Perform a sensitivity analysis. A sensitivity analysis will be quite useful in providing insights on MU risk. This focuses on assumptions related to interunit risk, including CCF and hazard correlation.

The importance analysis, sensitivity analysis and uncertainty analysis are discussed in Sections 4.6.1 to 4.6.5.

4.6.1. Review of results

The process described above provides a comprehensive approach that can be used as a basis for the MUPSA, with the intent to minimize both the underestimation and overestimation of MU risk. However, a review of the results needs to be performed to identify areas where the estimate of risk can be improved.

(a) **Review of quantification results.** The results of the MUPSA need to be reviewed using a similar process as the SUPSA. This would include steps discussed in IAEA-TECDOC-1804 [3].

(b) **Review of screening criteria.** Screening criteria, discussed in Section 4.2.1, need to be reviewed to determine that potentially risk significant sequences were not initially screened out. For example, if the screening criteria were established on the assumption that the overall MUCDF would be at least 10% of the SUCDF, the final results can be reviewed to ensure the assumption is valid. Additionally, the potential cumulative contribution to risk needs to be reviewed for all quantitatively screened out scenarios. For MUPSA, the review needs to ensure that criteria are applied such that the cumulative risk contribution of the screened scenarios is small for all POS combinations. This review needs to be documented to ensure that the screening applied was appropriate. The review of screening criteria may involve sensitivity analysis, as discussed in Section 4.6.3 below.

(c) **Review of results for asymmetry.** The effects of the asymmetry of the model resulting from the methodology presented can be assessed based on an examination of the results, including the cut sets. There may be cut sets missing when the MU core damage involves independent events at the affected unit. This is not the case where the core damage involves interunit CCF and/or correlation. The analyst will get the same cut set for Unit 1 followed by Unit 2 and for Unit 2 followed by Unit 1, and they are expected to minimize to one cut set when the results are merged. Cut sets missing because of the quantification approach (e.g. MET) may result in underestimation of risk. Therefore, it will be necessary to review the final cut sets to see the extent to which these asymmetrical cut sets contribute and to estimate the extent of the potential impact on the total risk of MU core damage. This can be done by adding together the contribution of these cut sets and multiplying that contribution by the number of combinations. For example, if it was a two-unit site, the MUCDF of these cut sets would be multiplied by two (to account for the fact that Unit 2 core damage followed by Unit 1 core damage was not modelled). This adjustment is expected to be more complicated for sites with three or more units.

4.6.2. Importance analysis

Importance analysis can be performed using similar methods to those used in the SUPSA. There is an issue with combining importance measures for different hazards (which may be analysed separately, as discussed above), which is also indicated as an open issue for a SUPSA (see more details in Ref. [5]).

For the MUPSA, importance measures can be derived for each quantified risk metric. As such, BEs would have different importance measures for each risk metric, including for both MUCDF and SUCDF.

As mentioned in Section 4.5.1, the limitations and assumptions in the modelling need to be understood when estimating the importance measures. The review of asymmetry can also be used when interpreting importance measures. For example, if there is a missing cut set involving an increased HFE (see the example in Section 4.6.1.3 above), the importance of the missing HFE can be estimated using the HFE importance for the other unit.

The importance measures for SU and MUPSA can be compared to identify differences, including BEs that may be more important for MUPSA than for SUPSA — including the reason for the increased importance. This may be due to adjustments made to the modelling (discussed in Section 4.4) or the scope of the MUPSA accident sequence modelling.

4.6.3. Sensitivity analysis

An aspect of importance is to compare the significance of systems affecting individual units with those that affect multiple units. If improvements to site safety are envisaged, this distinction can be a factor in what additional safety measures are necessary and to what extent they will improve safety. The objective of the sensitivity analysis is to check plausible alternative assumptions for uncertainties connected with specific modelling approaches and to analyse the potential level of impact on the risk metrics. Some of the sensitivity analyses discussed below are already included in the analysis steps above in consideration of whether detailed modelling is necessary — such as in the detailed CCF or HRA. If the sensitivity case shows a significant impact on the final results, more detailed modelling is suggested.

The following need to be considered for the sensitivity analysis (and were also exercised in the IAEA MUPSA case study — see Appendix II):

(a) Radioactive release from one of the units;
(b) CCFs;
(c) Human interactions;
(d) Administrative shutdowns;
(e) Seismic/hazard correlations;
(f) Screening criteria.

Each of the above, other than screening criteria (discussed in Section 4.6.1.2), is discussed in the following sections. Sensitivity analysis can include other cases, depending on the site specific contributions to the MUPSA risk metrics and major uncertainties in the MUPSA.

(a) **Radioactive releases from one of the units.** One of the elements of a MU risk assessment is the concept that multiple units in (or progressing towards) a core damage scenario can impact on operator actions at all units. To investigate this issue, a sensitivity case needs to be designed with a scenario where one unit is assumed to reach core damage before the other unit(s) and to have an early release from the containment while the other unit(s) are still not yet in a core damage condition. Sensitivity analysis can include the following assumptions:

    (i)   Assuming all local actions or actions outside the MCR are impacted by the radioactive release and are failed for all except the damaged unit;

    (ii)   Assuming actions occurring after core damage are impacted by the existence of the contamination, which does not prevent the action altogether but increases the HEPs as a result of elements such as increased execution time (e.g. due to slower actions in protective clothing or the need to look for alternative pathways), increased stress and other potential performance shaping factors.

(b) **CCFs.** This sensitivity case is intended to check the impact of CCF related assumptions by applying more accurate CCF modelling for several dominant CCF groups. CCF groups can be selected based on their importance contribution to MUCDF. An example of such sensitivity analysis, performed in the IAEA MUPSA case study, has shown that the CCF assumptions greatly impacted the MUCDF estimates for internal events and internal hazards. In the context of the case study, the total MUCDF was dominated by seismic MUCDF, hence modifying the simplified CCF modelling would have limited impact on the final results. This observation could not be generalized, since it might not be applicable for sites where the MUPSA results have significant contributions from internal events or internal hazards. In such cases, it is expected that detailed CCF modelling will have a significant impact on the final results.

(c) **Human interactions.** The analysis of operator actions in the MUPSA includes consideration for dependency, as well as the impact from radioactive release from another unit. Assumptions in the base model could be impacted by either the assumed timing or by other factors, such as shared resources. For the MUPSA, the dependency models used in the SUPSA can be very conservative, especially in the context of fully separate crews responding to each unit's event. The assumptions for interunit human dependencies need to be considered in sensitivity. In addition, the impact of timing needs to be reviewed where the potential exists for a time difference between core damage sequences at different units that would potentially result in the contamination of areas needed to perform operator actions attempting to prevent core damage. This potential is more likely for a site with three or more units, or with dissimilar units. A sensitivity analysis for this case can be performed by setting all local actions to an increased value.

(d) **Administrative shutdowns.** Assumption 2 provided in Section 3 states that administrative shutdown of the otherwise unaffected unit does not need to be modelled in the MUPSA. This assumption remains valid if the MUCDF or other risk metrics are not significantly below the SUCDF. However, if the MUPSA estimates the MUCDF risk to be several orders of magnitude below the SUCDF, then screening administrative shutdowns may in fact result in the screening of risk significant scenarios. In this case, a sensitivity analysis can be performed to include analysis of one or more scenarios that are important for the SUPSA, with an assumed administrative shutdown of the second or additional units.

(e) **Hazard correlations.** MU hazard correlations are expected to affect the final results. As mentioned in Section 4.4.5.3, the initial modelling of correlation typically assumes full correlation between similar components on similar units. The level of correlation is impacted by a number of factors, including the separation between the units, homogeneity of seismic hazards, orientation of the component and component design. The sensitivity analysis is intended to check the need for refinement of the fully correlated assumption if the results are impacted by modelling of partial correlations. This sensitivity analysis needs to determine what impact could occur if the correlation refinement is performed. The sensitivity can include a range of correlation, including assuming zero correlation as well as use of a 50% correlation factor. Additionally, this can include implementing these assumed correlations for all fragilities, as well as a smaller subset of risk significant fragilities (e.g. for ABWR MUPSA in the UK [28], the sensitivity analysis considered the top 50 risk-important fragilities in the seismic PSA).

(f) **Screening criteria.** The objective of this sensitivity analysis is to verify that potentially risk significant sequences were not initially screened out. The validity of the assumptions made to set up the screening criteria (e.g. the expected level of MUCDF) and the potential cumulative contribution to risk for all quantitatively screened out scenarios needs to be analysed. More details on the verification of screening criteria are provided in Section 4.6.1.

4.6.4. Uncertainty analysis

The uncertainty analysis for the MUPSA involves similar methodology and documentation to that for the SUPSA, but includes the MUPSA specific assumptions discussed above. Any additional model specific assumption needs to be documented and assessed as part of the MUPSA uncertainty analysis. Additionally, any model simplifications need to be documented and reviewed to determine the potential impact on the MUPSA results. Any simplifications determined to be potentially significant have to be subject to sensitivity analysis, assuming an alternate model can be easily quantified. Simplifications include any possible modelling of limited combinations, as discussed in Section 4.5.2, which can affect both the calculation of importance measures and the parametric uncertainty. Additionally, simplifications can impact the analysis performed in sensitivity studies, as discussed in Section 4.6.3.

As mentioned previously, it may be difficult to perform parametric uncertainty analysis, depending on whether the MUPSA results can be combined into a single set of results, or the simplification results in an estimate of risk metrics based on representative analysis (e.g. Unit 2 leading to Unit 1 CDF is estimated from an analysed Unit 1 sequence leading to Unit 2 CDF). However, a parametric uncertainty may be useful, even with limitations, to understand the range of the parametric uncertainty for each risk metric.

4.6.5. Interpretation of results

The results of the MUPSA can be used to improve overall plant safety, focusing on MU interactions and site level considerations. Interpretation of the MUPSA results involves the review of key insights, which include, but are not limited to, the following insights:

— Acceptability of MU risk results in comparison with safety goals (if any);
— Importance of particular SSCs in the MU context and comparison with the relevant SU risk profile (considering all permutations in MU risk metrics, e.g. two-unit MUCDF and four-unit MUCDF);
— Potential improvements in the robustness of shared systems and resources, which could have significant benefits from a MU risk perspective;
— Availability of human resources in case of MU accidents on-site;
— Sufficiency of emergency procedures and considerations of priorities in the case of MU accidents on-site.

Potential areas for MUPSA results application include various aspects of plant design safety, plant operation and oversight activities. The following items present examples of the application of MUPSA results and insights:

— Demonstration that site level safety goals are met, if such are established in the Member State;
— Design improvements, where MUPSA results could assist in focusing on MU interactions and shared systems;
— Improvement in plant emergency operating procedures, where the MUPSA results could potentially affect overall planning, including post-accident staffing requirements;
— Training of personnel in the response to MU accidents;
— Evaluation and setting up of priorities at the site in the case of shared resources;
— Regulatory applications, such as inspection planning and event analysis;
— Informing, if relevant, the resources needed to manage the emergency response if multiple units are simultaneously in emergency conditions (following Requirements 6 and 21 of GSR Part 7 [8]).

4.7. DOCUMENTATION OF THE RESULTS

SSG-3 [49] provides recommendations on the documentation requirements for a PSA, and IAEA Safety Reports Series No. 96 [1] and IAEA-TECDOC-1804 [2] provide additional guidance and information on MUPSA. Documentation of the MUPSA needs to be consistent with the documentation requirements for the SUPSA, including the detailed analysis supporting each of Sections 4.1 to 4.6, as applicable. If a scoping MUPSA approach is used, as discussed in Section 2.2.3, the documentation will include the base analysis as well as any insights, as discussed in Section 4.6. If a detailed MUPSA approach is used, the documentation will include the detailed steps of the analysis, as well as a detailed interpretation of the results in a similar fashion to the SUPSA. Documentation and information storage are performed in a similar way to a SUPSA in a manner facilitating peer review, as well as future upgrades and applications of the PSA, by describing the processes that were used and providing details of the methods applied, assumptions made and their bases.

When presenting results from a MUPSA involving site level risk metrics, it needs to be clarified in the documentation which sources are included (reactor core, SFP, etc.).

In general, the Level 1 MUPSA documentation[11] needs to include the following information:

(1) Conclusions and risk insights from the MUPSA, especially those that are different from those of the supporting SUPSAs;
(2) Guides and standards used to support the MUPSA, including which parts of these were included and excluded;
(3) The basis for the MUPSA approach selected and a description of its development;
(4) The basis for the selection of risk metrics;
(5) Documentation of specific assumptions used in the MUPSA;
(6) The process of model simplification, including a review of screened out IEs;
(7) The process and criteria for selecting POSs and POS combinations;
(8) Analysis of the IEs for their impact (e.g. single or MU), including analysis of internal and external hazards;
(9) Interunit data and common cause analysis;
(10) MU HRA and resulting HEPs and HEP dependencies;
(11) Quantification of any MU hazard correlation events;
(12) Quantification of each applicable MU risk metric and a summary of the results;
(13) Sensitivity analysis for the MUPSA;
(14) Identification of model uncertainty.

In addition, the documentation needs to include the creation of each unit's PSA, including the BE naming scheme, treatment for shared or interconnecting systems and analysis of any new data supporting the additional units' PSA.

5. LEVEL 2 AND LEVEL 3 MUPSA

As discussed earlier, MU risk aspects may extend from Level 1 to Level 2 and 3 PSA. While experience with Level 1 MUPSA is relatively limited worldwide, although growing, the extension into Level 2 and Level 3 PSA under current requirements is even more limited (nevertheless, some national regulations

---

[11] If the scope of the MUPSA includes Level 2 and/or Level 3, each documentation step should be performed for each scope, as applicable. Additionally, some documentation steps may require specific application for hazards within the scope of the MUPSA, such as internal hazards (e.g. fire or flood) or external hazards (e.g. seismic or high winds).

explicitly require the consideration of MU impacts in Level 2 PSA implementation). Although many legacy Level 3 PSAs were performed from the 1970s to the 1990s, most of these are very old and do not meet current industry PSA standards. While the IAEA case study did not explore a detailed assessment of Level 2 and 3 MUPSA aspects, and the focus of this safety report is on developing a technically robust basis for Level 1 MUPSA modelling, it is recognized that important questions remain that could benefit from advancing the MUPSA methodology described in this publication to Level 2 and 3. Some key technical issues raised in Section 1, such as the impact of the increased source terms and timing considerations from a MU release on early health effects, will need a Level 2 MUPSA. Hence, while this section is not as detailed as Section 4, the general principles of Level 2 and Level 3 MUPSA are presented. Detailed discussion of each step in the methodology is included in Sections 5.1 to 5.6. Additional information on the performance of a Level 3 MUPSA is available in Safety Reports Series No. 96 [1], based on early work done for Seabrook.

Figure 6 is a flow chart for the Level 2 MUPSA principles presented in this report, with the section numbers that provide the details of each step noted. As mentioned in Section 4, although the process shown in Fig. 6 is visualized as linear, the process performed for the MUPSA, as with SUPSAs, is iterative. For example, much of the work in each box can be performed in parallel, and if the initial results include application of conservative CCF or HRA, and the conservative approach greatly impacts the results, refinement of the CCF or HRA could be needed followed by reperformance of subsequent steps.

The steps to perform the Level 2 MUPSA would be similar to those discussed in Section 4. However, some additional considerations are needed to address the specific steps and modelling aspects for a Level 2 PSA. SSG-4 [50] includes an overall approach for a Level 2 PSA, including the high level aspects of the PSA reviewed below for impacts due to MUPSA.

The following sections summarize the additional considerations for the steps in Section 4, considering Level 2 modelling factors. Details of the Level 1–Level 2 interface, accident progression analysis and source term analysis are discussed in Sections 5.4.1, 5.4.2 and 5.4.3. The methodology used to solve the Level 2 (or Level 3) PSA may be impacted by the methodology used to solve the Level 1 MUPSA, which is discussed in Section 4. It may be that the base SUPSA methodology for Level 2 may need to be adjusted to solve the Level 2 MUPSA.

As discussed above, the base PSA model (including the Level 2 model) needs to be reviewed to ensure the SUPSA model adequately addresses the MU effects in the base model. This includes considerations such as shared resources or control room, common site mitigation provisions, interunit connections and possible accident propagation between units. As a starting point for the Level 2 MUPSA modelling, the issues discussed in Section 4.2 need to be reviewed for the SUPSA.

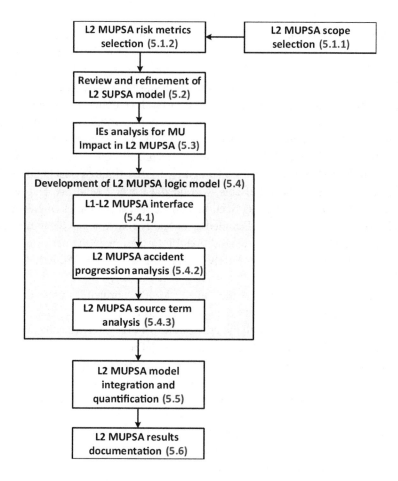

FIG. 6. Level 2 (L2) MUPSA methodology.

5.1. LEVEL 2 MUPSA SCOPE AND RISK METRICS SELECTION

5.1.1. Scope selection

The MUPSA scope begins with the selection of hazards and POSs that are risk significant and those involving the potential for MU impacts. When including Level 2 (or Level 3) PSA, the risk significance consideration, as well as the review of MU impacts, would include consideration of Level 2 (or Level 3) risk significance and Level 2 PSA model impacts.

5.1.2. Risk metrics selection

Risk metrics associated with Level 2 MUPSA are discussed in Sections 2.3 and 4.1.2. Additional risk metrics would be selected if the SUPSA involved Level 2 analysis or analysis of shutdown or SFP risk. The base Level 2 risk metrics typically include site LRF, SULRF, site LERF, SU LERF, SU RC frequency and MURCF.

The selection of MURCF may involve additional Level 2 analysis to determine the release impact of unique combinations of releases from each unit, especially if the RC results are going to be used for Level 3 PSA or if simplification of the MURCF is performed (e.g. additional grouping beyond the SUPSA grouping).

Note that not all of the applicable Level 1 or Level 2 risk metrics need to be selected and analysed. The case study demonstrated that the risk metric providing the most insights involves MU sequences (e.g. MUCDF, MULRF or MURCF).

Other risk metrics may be selected, as discussed in Section 2.4, depending on the use of the MUPSA and whether the site includes more than two units. For example, if the site contained two pairs of two similar units (four units total), the Level 1 risk metrics might involve MULRF for the two sets of two-unit pairs, plus a MULRF for all four units. Finally, the results would also likely involve the calculation of the ratio of MU to SU risk metrics. This MU ratio was calculated in the IAEA case study discussed in Appendix II.

The basis for the selection of risk metrics needs to be documented, as discussed in Sections 4.7 and 5.6.

5.2. REVIEW AND REFINEMENT OF LEVEL 2 SUPSA MODEL

The steps discussed in Section 4.2 for simplifying and refining the SUPSA are generally the same when including the Level 2 (or Level 3) PSA in the scope of the MUPSA. If Level 2 or Level 3 end states are included in the MUPSA, the following model refinement steps for Level 1, Level 2 and Level 3 are performed simultaneously:

(1) **Level 2 SUPSA model simplification:** to simplify the Level 2 SUPSA model, by screening low risk contributors to the risk metrics previously selected;
(2) **Level 2 SUPSA model refinement:** to refine the Level 2 SUPSA model as needed to ensure that the MU accident sequences are correctly modelled;
(3) **Level 2 SUPSA models development for all units:** to create an individual Level 2 PSA model for each unit to be considered in MUPSA.

In Step 1, the screening discussed in Section 4.2.1 would be expanded to include the risk significance of the IEs for the SU Level 2 (or Level 3) PSA. In Step 2, the refinement of the model is complicated by the expansion of this modelling to include Level 2 accident sequences. During this refinement, low risk accident sequences can be removed (screened out) from the model to simplify the MUPSA. This refinement would include screening of the resulting Level 2 sequences. Finally, Step 3 would include the creation of Level 2 (or Level 3) PSA models for all units on-site within the scope of the MUPSA.

5.3. IE ANALYSIS FOR MU IMPACTS

The step for IE analysis is not greatly impacted by the inclusion of Level 2 (or Level 3) PSA in the MUPSA. Additional consideration is needed to check whether the IE might cause a potential degradation on another unit. With this step, the potential degradation includes consideration of the systems discussed in Section 5.4. Degraded Level 2 systems affected by the IE include containment systems and would be impacted if the units share containment or associated systems. The IE documentation step needs to include the consideration for Level 2 systems potentially impacted by MU accident sequences.

5.4. LEVEL 2 MUPSA LOGIC MODEL DEVELOPMENT

The inclusion of the Level 2 logic model in the MUPSA can add considerable complication to the modelling. Since every SUPSA Level 1 accident sequence results in multiple Level 2 accident sequences, and the modelling of combined MU accident sequences from multiple units multiplies the number of resulting MU sequences over what is in the SUPSA, it is easy to see that the number of resulting MU Level 2 accident sequences can become quite large. The number of resulting MU plant damage states (PDSs) or MU RCs would be significantly higher than the number for the SUPSA. For example, if a SUPSA included 20 RCs, the MUPSA could potentially have 400 MU RCs. In practicality, the number is smaller, but still in the order of 100. Model simplification, including both IE screening and accident sequence screening, can reduce overall the resulting number of MU RCs. For RCs large enough — either individually or in combination — that early fatalities (deterministic consequences) may be significant, the relative timing of the releases from each unit may also need to be considered. Simultaneous releases may have different off-site consequences in terms of the number of early fatalities than releases offset in time, as a result of changes in meteorological conditions exposing more people and the non-linear risk–dose curve (see Annex V).

In developing the MUPSA modelling for Level 2, the containment event tree logic can become considerably more complicated than the logic discussed for Level 1 in Section 4.4.2. However, the logic can more accurately model the impact of releases causing dose impacts to other units on-site, as discussed in Section 4.4.4. For example, for a two-unit site, if Unit 1 containment is intact, the impact on Unit 2 following Unit 1 core damage may be minimal in comparison to the impact on Unit 2 if the Unit 1 containment failed. Because of the complicated logic for the MUPSA resulting from adding the Level 2 containment events to the MUPSA model, the MET approach could be more difficult to apply than the SFT approach. However, either of the approaches discussed in Section 4.4.2 can be used in the Level 2 MUPSA.

An important insight from the Seabrook Level 3 MUPSA discussed in Safety Reports Series No. 96 [1] is that when a MU accident is initiated by a seismic event or SBO, the accident progression sequences through the containment event tree are focused on a very limited number of applicable PDSs and containment event tree sequences.

An example of developing a Level 2 PSA logic model is provided in Annex IV, Section IV–1.1.1. The example analysis includes modelling of PDSs and RCs extended to MU sites.

The system modelling, HRA and data analysis discussed in Sections 4.4.3, 4.4.4 and 4.4.5 are the same for Level 1 and Level 2 MUPSA. New HEPs and dependencies would be included in the Level 2 PSA, including actions directed by the severe accident management (SAM) guidelines. Sections 5.4.1 and 5.4.2 below discuss some specific considerations for the Level 1–Level 2 interface and the phenomena associated with the accident progression analysis.

For the HRA, the radiological release impact is more accurately assessed when Level 2 is included in the scope of the MUPSA, since the release from one unit is assessed on a sequence-by-sequence level. For example, when only considering Level 1 CDF for a two-unit site, if Unit 1 experienced core damage, the dose impact on Unit 2 would have to be assumed. When including the Level 2 accident sequences, each resulting Level 2 sequence on Unit 1 would be assessed with a different dose release. This allows for a more precise assessment of the HRA considerations for dose.

5.4.1. Level 1–Level 2 interface

The Level 2 model, which involves mitigation of severe accidents, including support systems, operator actions and containment integrity, can be modelled in a similar way to the Level 1 PSA modelling when accounting for MU aspects.

The assessment of PDSs for the Level 1–Level 2 interface is not expected to be impacted by the MUPSA methodology. However, this assumption cannot be extended to other reactor designs, where the units have significant relevant shared systems. For example, the challenges to containment integrity depend on the number of accident units, and the PDS development is based on whether the accident involves one reactor unit or more than one reactor unit.

Both the IAEA case study and the ABWR pilot evaluation in the UK showed that a vast majority of the MUPSA CDF sequences involving similar units involved both units experiencing the same PDS. This information is useful in determining the relative timing for core damage, as well as the assignment of MU RCs discussed below. However, even with similar units experiencing the same PDS, this does not guarantee the same timing for releases. For example, with an SBO PDS involving EDGs failing to run, the run failure could occur at any time during the mission time.

5.4.2. Accident progression analysis

The major issue not already covered under the Level 1 methodology above involves correlation between Level 2 phenomenological factors affecting Level 2 accident progression. These factors can include, but are not limited to, the following:

— Conditional containment failure probability;
— Conditional steam explosion (in-vessel or ex-vessel);
— Hydrogen detonation;
— High pressure core melt ejection;
— Conditional pipe break probabilities (important for interfacing system LOCA);
— Molten core–concrete interaction.

One approach for addressing these Level 2 MUPSA factors is similar to the approach for CCF or correlation in Section 4: to initially model these factors conservatively and then refine the model for risk significant factors. In this case, the refinement would be to review the factors in more detail and develop a qualitative argument as to why the factor is or is not correlated between units.

5.4.3. Source term analysis

The assessment of source terms, and the resulting RC assignments, is fundamentally unchanged for the MUPSA. The existing RC grouping used by the SUPSA can be used for the MUPSA. The RCs provide information on the timing, size and location of the release, which can be useful in the assessment of the impact of core damage from one unit affecting an adjacent unit or other unit on-site.

If the RCs are being used as an input to the Level 3 PSA, it may be useful to develop simplified groupings for MU RCs. This can help simplify the Level 3 analysis. The initial output from the Level 2 MUPSA would involve sequences where Unit 1 experiences one PDS and RC, while Unit 2 experiences another PDS and RC. As mentioned above, the ABWR evaluation in the UK [28] showed that most of the time, the same PDS and RC were experienced by both units, although this is not always the case. As a practical matter, a MU RC can then replace the RCs assigned to each unit.

Release timing can affect the RC categorization as well the impact on the HRA (dose impact on the adjacent units). Analysis of the release timing, and the impact on off-site dose, is discussed in Annex IV. Section 5.5.1 discusses the results of the IAEA case study and the UK ABWR results, which concluded that similar units would experience similar PDSs a vast majority of the time, but that even within the same PDS, some timing differences are possible.

As an example, assume that a two-unit site has 30 RCs for the SUPSA. The most risk significant sequences involve RCs 14 and 16 (in this example). When analysed, the MUPSA will show four combinations, with Unit 1 and Unit 2 experiencing either RC14 or RC16. Let us assume for this case that RC14 has roughly a factor of ten higher release than RC16. As a simplification, the MUPSA can reduce the U1-RC14/U2-RC16 and U1-RC16/U2-RC14 combinations to a single MU RC, since the source term off-site looks the same. In fact, the releases off-site for the four combinations may be further simplified to two RCs, since the difference between both units experiencing RC14 and one experiencing RC14 with the other experiencing RC16 is only a factor of two.

This approach would be needed, since in theory, the 30 RCs in this example could end up with some 900 possible combinations for Unit 1 and Unit 2 RCs. Without simplification, the Level 3 analysis (if performed) could be difficult to manage. As a result, a simplification of the MU RCs is suggested. Figure 7 shows another example of a modelling approach applied for nine units in the Republic of Korea, by simplifying the RCs into five groups as follows: (1) bypass, (2) not isolated, (3) early containment failure (leak and rupture), (4) late containment failure (leak and rupture) and (5) basemat melt through.

RCs for either shutdown POS or SFP core damage sequences could also complicate the MUPSA analysis. However, the general approach is similar, whether considering the impact of core damage on an adjacent unit or solving the Level 2 model for the resulting RC combinations.

A review of RC combinations, extending to Level 3 release estimates, is provided in Annex V. Research performed in the UK concluded that an additive

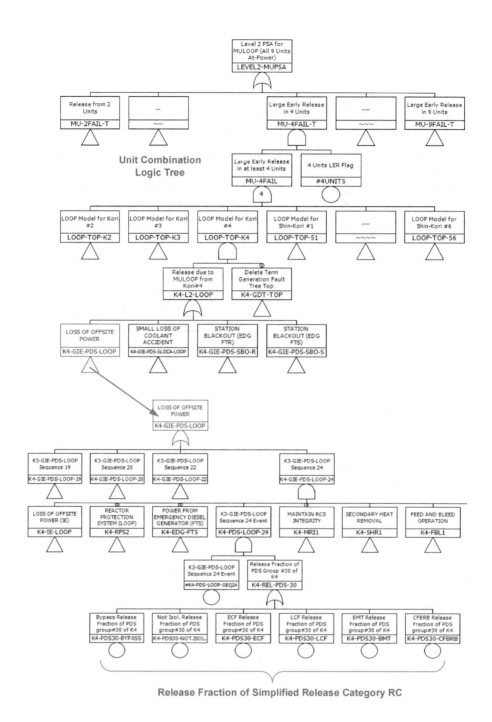

FIG. 7. KHNP's modelling approach to Level 2 MUPSA.

combination (for individual risk) or linear combination (for societal risk) of SU results provided a good approximation for simultaneous MU releases. However, more complex effects were seen for time offset releases and some areas of potential non-conservatism for deterministic consequences.

5.5. LEVEL 2 MUPSA MODEL INTEGRATION AND QUANTIFICATION

The model integration process and MUPSA quantification is similar to the process discussed in Section 4.5 when including Level 2 in the MUPSA. The number of overall accident sequences and uncertainties affecting the MUPSA will result in a more complicated but complete analysis. The analysis includes the expansion of the MUPSA quantification to include both the Level 1 and Level 2 risk metrics, as discussed in Section 5.1.2. This includes an expansion of sensitivities discussed in Section 4.6.3. The existing sensitivity groups discussed in this section are a good starting point, but the issues being analysed would be expanded to include additional Level 2 (or Level 3) considerations. The issues impacting the analysis are discussed in Sections 5.4.1, 5.4.2 and 5.4.3. Additional discussions on Level 2 and Level 3 sensitivities are provided in Annex V.

5.6. DOCUMENTATION OF LEVEL 2 MUPSA RESULTS

Section 4.7 includes a list of 11 areas to document in the MUPSA. The documentation for the Level 2 MUPSA includes the same 11 areas but needs to be expanded to include the specific Level 2 considerations discussed in Section 5.4.

In particular, the documentation needs to include the following considerations:

(a) The timing of releases for units experiencing both the same and different RCs;
(b) The modelling of phenomena affecting the Level 2 MUPSA model (see Section 5.3);
(c) The uncertainty related to the source term analysis, including timing of the releases and effects or RC analysis simplification.

5.7. LEVEL 3 MUPSA

The advantage of performing a Level 3 MUPSA is that site risk metrics such as individual dose, individual risk or societal risk provide a direct measure of the risk to people and the environment and can be compared with any site public risk goals, if applicable (see Section 2.3.3). These may already exist or can be developed. This also avoids the need to define what constitutes 'large' or 'early' in the Level 2 LRF and LERF metrics if only a Level 2 PSA is to be performed.

Level 3 MUPSA involves calculation of the total risk from all units at the site. Some issues to consider when defining risk metrics or performing calculations for individual and societal risk, respectively, are outlined in the following list:

(a) Individual risk considerations
    (i)   The representative person — the individual for whom the risk from the site is calculated — may need careful selection, taking account of the locations of the individual units with respect to potential exposed individuals. A simpler approach may be to assume that all releases occur from a single point, but this may lead to an overestimate of the risk.
    (ii) Decisions on pathways to include, whether and which countermeasures (protective actions and/or remedial actions) can be credited and integration times for the deposited dose will also need to be made.
(b) Societal risk considerations
    (i)   Aggregating low doses over large numbers of people could lead to conservative results with large numbers of notional fatalities if the linear no threshold assumption is applied.
    (ii) Certain judgements will need to be made on the geographical and temporal extent over which to perform the calculations.
    (iii) Consideration of the demographic situation and the tendency of population density to increase.

Dose assessment for a MUPSA release could be performed in a similar way to SU Level 3 PSA principles, based on the Level 2 MUPSA results and the MU RCs assigned to each MU accident sequence. Annex V discusses one approach to performing this dose assessment; and Annex VI discusses related EPRI work, including the factors affecting the analysis and the timing of the release from each unit. The software used for off-site dose estimation needs to include multipoint release when considering unique release points or timing, although single point release models can approximate the off-site dose in most cases.

Studies in the USA have shown that multiple releases have generally led to a proportionate increase in consequences (see Annex VI). However, a recent study in the UK (see Annex V) has shown that for RCs large enough — either individually or in combination — early fatalities may be significant, and the relative timing of the releases from each unit may also need to be considered. Simultaneous releases may give different off-site consequences in terms of the number of early fatalities to those from releases offset in time; this is a result of changes in meteorological conditions — principally wind direction — during the releases exposing different numbers of people at different levels and the non-linear risk–dose curve for deterministic effects. In these cases, the results are likely to be very sensitive to the exact population distribution and changes in meteorological conditions.

As mentioned above, if a Level 3 PSA is performed for MUPSA, many more RCs may need to be analysed, and for combination RCs, different offsets in timing may also need to be examined for large releases. Further discussion on initial Level 3 analysis for MUPSA is provided in Annexes V and VI, including the referenced ONR research.

6. PATH FORWARD FOR MUPSA

This publication provides technical details for screening site specific internal and external hazards, including combinations thereof, and analysing the overall site risk related to the MU risk of core damage, large release and other risk metrics. Although the principles and approaches provided in this safety report have been successfully applied to both the IAEA case study and other MUPSA pilots, additional advances are needed to fill gaps in the methods and data supporting a comprehensive MUPSA.

Further sections describe the areas where further development and investigation is needed to support the methodology described in this publication. The areas and challenges described below are based on the experience gained during the MUPSA project, including extensive interaction and technical discussions held during consultancy and technical meetings with PSA practitioners from various Member States.

6.1. COMPLEXITY AND SIZE OF MUPSA MODEL

Development and quantification of MUPSA risk metrics entails several technical challenges. The model size for MUPSA models can challenge PSA software, including the solution algorithms. Current SUPSA models can already be quite large and are difficult to analyse, especially when trying to obtain convergence of the models. Convergence is a more complex issue in the context of analysing of two or more units (see examples in Annex IV).

Thus, efforts might be necessary to improve and optimize the solution algorithms and to explore the potential for focusing the scope of MUPSA studies on risk significant areas. While the supporting case studies have demonstrated the feasibility of performing a MUPSA on a site with two pairs of identical units, more experience is needed to demonstrate the applicability of this methodology to larger and more complicated sets of reactors and shared facilities. In addition, experience in modelling and evaluating accidents involving non-core sources and combinations with one or more reactor sources is lacking.

6.2. SMALL MODULAR AND ADVANCED REACTORS

Small modular reactors and advanced reactors often have unique designs, such as multimodules feeding a single turbine generator and multimodules in a single large pool. The application of MUPSA to small modular reactors or advanced reactors has not been extensively analysed beyond initial studies [61, 62]. The Power Reactor Inherently Safe Module PSA included a pilot application of the ASME/ANS non-LWR standard [17], which included requirements for MUPSA [62]. It is expected that the MUPSA methodology will be applicable for specific small modular reactor designs, including multimodule considerations. However, the methodology presented in this safety report does not cover explicitly the potential challenges and specifics of multimodule risk assessment for small modular reactors. In addition, non-LWRs typically perform Level 3 PSAs, since core damage has not been found to be a good surrogate of public risk for many non-LWR concepts.

Thus, efforts might be necessary to supplement the methodology with the specifics of small modular reactors supported by the case study and considering various designs (including non-LWRs and/or advanced reactors).

6.3. CCF ANALYSIS

One of the challenges in MUPSA is the data to support MU CCF analysis (including the applicability of CCF data across units) [29]. There have been a significant number of recent studies in the area of MU CCF, as noted in Section 4.4.5.2. This includes the development of both simplified and detailed modelling for sites with many reactors. Some of this effort is performed to minimize the amount of effort or analysis time for the MUPSA. However, currently, the guidance in this area is not extensive.

A particular area of interest in this context is the analysis of MU CCF for passive and software based systems. These aspects are challenging also from a SU perspective and amplify in the context of MU sites.

Thus, further efforts are needed to collect and process the data for MU CCF and elaborate on the methods for CCF of passive and software based systems.

6.4. MU HRA ISSUES

Current HRA dependency models are based on HFEs performed in the same unit, typically by the same crew using the same procedure, which is not fully covering the potential challenges in the MU context (when more than one crew is involved). Some of the factors important to SU dependency modelling (e.g. training, common procedures, management) are applicable across units, whereas other factors (e.g. same crew, timing, environment) are not (or may not be) applicable. Also, the effect of radiological releases or degraded conditions can be analysed using existing HRA methods, as demonstrated in the IAEA case study. One issue, however, is the impact of release timing. Uncertainty related to timing can have a significant impact on the HRA, mainly on local human actions. In addition, as discussed in Section 4.4.5, MU events can be impacted by organizational factors, which was shown to be an influencing factor during the events at the Fukushima Daiichi NPP. However, current HRA methods have difficulty assessing the direct impact on organization factors in a SUPSA, which translates into difficulty in assessing the impact for a MUPSA.

Thus, further efforts are needed in the area of improving the models of interunit HFE dependencies and understanding the level of impact of MU accident conditions on contextual characteristics (e.g. performance shaping factors). Additionally, further research is needed to determine if and how the HRA can be affected by MU impacts related to organization factors.

6.5. LEVEL 2 MUPSA FACTORS

As noted in Section 5, some national regulations explicitly require consideration of MU impacts in Level 2 PSA (e.g. Ref. [63]), but still the experience with Level 2 MUPSA is limited worldwide. As also noted, the MUPSA approach presented in this safety report is generally applicable to Level 2 and Level 3 PSAs. However, the factors listed in Section 5.4.2 are identified as areas needing further study to determine whether they are correlated among the units and cannot be treated as independent factors. These factors include:

— Conditional containment failure probability;
— Conditional steam explosion (in-vessel or ex-vessel);
— Hydrogen detonation;
— High pressure core melt ejection;
— Conditional pipe break probabilities (important for interfacing system LOCA);
— Molten core–concrete interaction.

The methodology discussed in Section 5.4 proposes a conservative treatment for these factors initially. However, when the factor becomes significant, more detailed analysis is suggested. Thus, further research in this area is needed to support consensus modelling of each factor.

6.6. MULTIPLE POS

There are currently no examples of full application of the MUPSA approach to plants with multiple POSs. This analysis complicates an already complicated PSA model, starting with the IE analysis, extending to the CCF modelling and the Level 2 analysis. Additionally, the combined source term analyses performed as example studies in Annexes V and VI were performed for units at full power, which does not consider the releases that occur from multiple sources both at the same site and at different sites.

In general, it is expected that the application of the approach discussed above and supplemented with the analysis approaches in the annexes can be applied to a MUPSA considering all POSs. However, further efforts on a detailed MUPSA for multiple POSs are needed to fill this gap.

Appendix I

LEVEL OF CHANGES NECESSARY FOR
TYPICAL PSA TASKS IN MU CONTEXT

One of the main assumptions for MUPSA implementation is the availability of SUPSA as a prerequisite. SUPSA models are assumed to be of sufficient quality and to be implemented using the state of practice in the area of PSA.

As stated in the objectives of this report, the MUPSA methodology is built on the basis of the current state of practice in PSA and assumes implementation of the typical PSA tasks in the MU context.

Table 2 shows the typical tasks of Level 1 PSA in accordance with SSG-3 [49] and coded in compliance with IAEA-TECDOC-1804 [3]. The level of changes necessary for typical PSA tasks for MUPSA in comparison with traditional SUPSA practices was evaluated based on the engineering judgement in Phase I of the MUPSA project and was later revised based on the experience gained during the case study. Table 2 presents the ranking of the typical tasks of Level 1 PSA in terms of the level of changes necessary to address the MU context. The following four categories were used for ranking:

— Category I: Task could add significant changes and complexity in comparison with the SUPSA.
— Category II: Task could add moderate changes and complexity in comparison with the SUPSA.
— Category III: Task could add minor changes and complexity in comparison with the SUPSA.
— Category IV: Task does not add supplementary changes and complexity in comparison with the SUPSA.

The ranking category assigned to each PSA task is based on engineering judgement and is intended to provide a general understanding of the level of effort necessary for MUPSA.

TABLE 2. RANKING OF THE TYPICAL PSA TASKS IN TERMS OF LEVEL OF CHANGES NECESSARY FOR MU CONTEXT

PSA tasks as per IAEA-TECDOC-1804 [3]		Category
Plant operational state (OS) analysis		
OS-A	Identification of POSs	II
OS-B	POS grouping	III
OS-C	Estimation of POS frequencies and durations	II
OS-D	Documentation	III
Hazard events (HE) analysis		
HE-A	Identification of potential hazards	III
HE-B	Hazard screening and final hazard list identification	II
HE-C	Characterization of hazard events for all hazards	IV
HE-D	Characterization of hazard events for internal fires	IV
HE-E	Characterization of hazard events for internal floods	IV
HE-F	Characterization of hazard events for seismic hazards	IV
HE-G	Frequency of hazard events for all hazards	IV
HE-H	Frequency of hazard events for internal fires	III
HE-I	Frequency of hazard events for internal floods	III
HE-J	Frequency of hazard events for seismic hazards	III
HE-K	Fire scenario development	III
HE-L	Flood scenario development	III
HE-M	Documentation	IV

TABLE 2. RANKING OF THE TYPICAL PSA TASKS IN TERMS OF
LEVEL OF CHANGES NECESSARY FOR MU CONTEXT (cont.)

PSA tasks as per IAEA-TECDOC-1804 [3]		Category
	Initiating event (IE) analysis	
IE-A	Identification of IE candidates (preliminary identification of IEs)	II
IE-B	IE screening and final IE list identification	II
IE-C	IE grouping	III
IE-D	Collection and evaluation of generic information for IE frequency assessment	III
IE-E	Collection of plant specific information	III
IE-F	IE frequency quantification	III
IE-G	Documentation	IV
	Accident sequence (AS) analysis	
AS-A	Selection of a method and provision of related tools for accident sequence modelling	II
AS-B	Definition of success and non-success end states and key safety functions	II
AS-C	Accident sequence progression identification and model development	III
AS-D	Accident sequence success criteria definition	III
AS-E	Documentation	IV
	Success criteria (SC) formulation and supporting analysis	
SC-A	Definition of overall and detailed success criteria	III
SC-B	Thermal hydraulic analyses and other assessment means supporting the derivation of detailed success criteria	III

TABLE 2. RANKING OF THE TYPICAL PSA TASKS IN TERMS OF
LEVEL OF CHANGES NECESSARY FOR MU CONTEXT (cont.)

PSA tasks as per IAEA-TECDOC-1804 [3]		Category
Success criteria (SC) formulation and supporting analysis		
SC-C	Documentation	IV
Systems (SY) analysis		
SY-A	System characterization and system boundary definition	II
SY-B	Failure cause identification and modelling	III
SY-C	Identification and modelling of dependencies	I
SY-D	Documentation	IV
Human reliability (HR) analysis		
Pre-IE HRA		
HR-A	Identification of routine activities	III
HR-B	Screening of activities	III
HR-C	Definition of pre-initiator HFEs	III
HR-D	Assessment of probabilities of pre-initiator HFEs	III
Post-IE HRA		
HR-E	Identification of post-initiator operator responses	II
HR-F	Definition of post-initiator HFEs	II
HR-G	Assessment of probabilities of post-initiator HFEs	II
HR-H	Recovery actions	II

TABLE 2. RANKING OF THE TYPICAL PSA TASKS IN TERMS OF
LEVEL OF CHANGES NECESSARY FOR MU CONTEXT (cont.)

PSA tasks as per IAEA-TECDOC-1804 [3]		Category
Human induced IE human reliability analysis		
HR-I	Identification of human failure events that could lead to an IE	III
HR-J	Grouping of human failure events that could lead to an IE	III
HR-K	Assessment of frequencies of human failure events that could lead to an IE	III
Documentation		
HR-L	Documentation	IV
Data analysis (DA)		
DA-A	Reliability model parameter identification	IV
DA-B	Component grouping for parameter estimation	III
DA-C	Collecting and evaluating generic information	III
DA-D	Plant specific data collection and evaluation	IV
DA-E	Derivation of plant specific parameters, integration of generic and plant specific information	IV
DA-F	Derivation of plant specific parameters for common cause failure events	I
DA-G	Use of mechanistic models (fragility analysis)	IV
DA-H	Documentation	IV
Dependent failures (DF) analysis		
DF-A	Design related dependency analysis	III
DF-B	Operations related dependency analysis	II

TABLE 2. RANKING OF THE TYPICAL PSA TASKS IN TERMS OF
LEVEL OF CHANGES NECESSARY FOR MU CONTEXT (cont.)

PSA tasks as per IAEA-TECDOC-1804 [3]		Category
Dependent failures (DF) analysis		
DF-C	Physical dependency analysis	I
DF-D	Common cause IE analysis	II
DF-E	Common cause failure analysis	I
DF-F	Subtle interactions	III
DF-G	Documentation	III
Model integration and risk metric frequency quantification (MQ)		
MQ-A	Integrated model	II
MQ-B	Requirements on the quantification	II
MQ-C	Review and modification of the results	I
MQ-D	Documentation	IV
Results analysis and interpretation (RI)		
RI-A	Identification of significant contributors	I
RI-B	Assessment of assumptions	I
RI-C	Documentation	III

Appendix II

MUPSA CASE STUDY

II.1. GENERAL DESCRIPTION OF THE CASE STUDY APPROACH

Phase I of this project resulted in a working material document providing a methodology for the implementation of MUPSA with practical PSA modelling tips. Phase II of this project was aimed at developing a case study following the methodology developed in Phase I, with the ultimate objective to improve the methodology based on the feedback from the case study. Thus, the case study was an essential part of the MUPSA project aimed at completing the efforts implemented in Phase I and supporting finalization of the project in Phase III (issuance of this safety report).

The objective of the case study is to verify the proposed MUPSA methodology by applying it to a realistic NPP configuration using a realistic PSA model and to provide feedback on the applicability of the proposed methodology for standard PSA tasks. In addition, the case study is expected to provide a base for improvement and increase in the level of detail reflected in the methodology.

II.1.1. General description of the case study

A realistic PSA model for a PWR type reactor was used for the case study. It was provided by RELKO Ltd (Slovakia). The case study was implemented by the steering group, which included the participants from the IAEA/NSNI MUPSA working group. The modelling task was conducted by Mr. Pavol Hlaváč from RELKO Ltd.

Based on the available SUPSA model, the site considered for the case study has been defined as a site with four PWR type units (Units 1 to 4) — Unit 1 and Unit 2 being of relatively old design (old units), and Unit 3 and Unit 4 being of the new design ('new' units). The layout of the site is illustrated in Fig. 8. The unit pairs were assumed to be identical. Various changes were made to the base plant design to create these old and new units in order to support the goals of the case study, so while these plants do not exactly represent any particular plant, they are a reasonable representation of the features that might exist at plants of similar design, but different vintages. The SU model was applied to a dual-unit site, expanded to represent a four-unit site (not an actual or existing site). Some of the unit differences were assumed with the explicit purpose of testing different elements of the MUPSA.

The old units (Units 1 and 2) were assumed to share the following features:

(a) Areas: turbine hall, specific cable tunnels and emergency control room;
(b) Systems: service water system, residual heat removal system, circulating water system, fire suppression water system, electrical switchyard and one mobile source for emergency feedwater of Unit 1 and Unit 2.

The new units (Units 3 and 4) were assumed to share the following systems: service water systems, circulating water system, SAM diesel generator, SAM emergency water source and one mobile source for emergency feedwater of Unit 3 and Unit 4.

All four units have the following features:

(a) Common external electrical grid;
(b) Shared portable systems to supply auxiliary water;
(c) Common recovery procedures;
(d) Located at the same elevation;
(e) Common maintenance staff and maintenance procedures;
(f) Same supplier for diesel fuel (however, diesel fuel supply lines are independent).

Further description of the case study units is provided in the case study report discussed in Appendix II.

II.2. SCOPE OF THE CASE STUDY

The case study covered analysis of examples of different type of IEs, such as internal events, fires and seismic events. The details of these areas of analysis are discussed in Sections II.2.1 to II.2.8.

The risk metrics considered in the case study are limited to consideration of MU core damage (core damage on two or more reactor units). SU core damage has also been considered for comparison and for identification of the MU ratio[12].

_____

[12] The MU ratio represents the ratio between SU and MU CDFs. This is essentially equivalent to the conditional probability of a second or subsequent unit going to core damage if one unit goes to core damage.

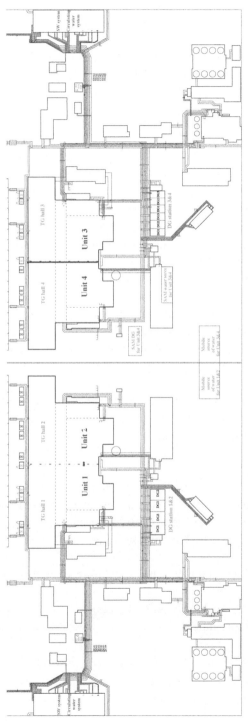

FIG. 8. Layout of the site analysed in the case study (TG: turbine generator; DG: diesel generator; SW: service water) (courtesy of Pavol Hlavac, RELKO).

II.2.1. IEs for the case study

At the time of the occurrence of the IE, all four units are assumed to be in full power operational mode. The screening of internal IEs resulted in the identification of the following list of non-screened MU internal IEs:

(a) Large condensate system rupture at Unit 1 or Unit 2;
(b) Feedwater line break at Unit 1 or Unit 2;
(c) Steam line break downstream of the steam isolation valve at Unit 1 or Unit 2;
(d) Steam line break upstream of the steam isolation valve (outside confinement) at Unit 1 or Unit 2;
(e) LOOP (common for Units 1–4);
(f) Loss of circulating cooling water system of Unit 1 and Unit 2;
(g) Loss of circulating cooling water system of Unit 3 and Unit 4;
(h) Loss of service water system of Unit 1 and Unit 2;
(i) Loss of service water system of Unit 3 and Unit 4.

Since the general approach to the analysis of each of these IEs would be the same, it was decided that it would be sufficient to perform detailed analysis on two of them to test the methodology. From the above mentioned events, the following internal IEs were chosen and have been analysed within the case study:

(a) Steam line break outside containment (SLBO) at Unit 1 with consequential loss of main feedwater (LMF) pumps at Unit 2 (in case of steaming in the turbine hall due to unsuccessful steam isolation);
(b) LOOP at all four units.

In addition to the internal IEs, testing the methodology required performing a detailed analysis for some hazard induced events. Again, the general approach would be the same regardless of the hazard, so three hazard induced events (one internal and two external) have been considered in the case study:

(1) Fire in the turbine generator hall of old Units 1 and 2, with fire spreading to the switchgear room of Unit 2;
(2) Seismically induced LOCA (small LOCA) at all four units;
(3) Seismically induced SBO at all four units.

The seismic hazard curve was decomposed into ten intervals of approximately 0.1g, from 0.05g to 1.1g.

II.2.2. Accident sequence analysis

Accident sequence analysis is performed using event trees. The MUPSA model contains SU event trees for selected IEs for all four units. Therefore, the SU risk for selected IEs can also be quantified within the MUPSA model.

The MU accident sequences are analysed using integrated event trees for two units or four units, depending on the IE. The integrated event trees use the same logic for each unit as the corresponding SU event tree for the corresponding IE. Furthermore, the SU event tree logics are integrated in order to develop MU event trees for selected IEs. For all selected IEs, the MET approach has been applied for integrated MU event tree development. The method is based on combining the event trees from different units and is described in detail in section 3.5.2 of the working material on the Phase I MUPSA methodology [54].

For the LOOP IE, both MET and single-top fault tree approaches have been applied. The single-top fault tree approach involves conversion of the accident sequence models for individual units into the equivalent fault trees logic, which are then used as functional events in the MU event tree. The logic represents all possible combinations of core damages at different units.

II.2.3. System analysis

The system reliability models used for SU event trees are also used for integrated MU event trees. If the SU system fault tree is affected by the MU condition, a boundary condition set has been introduced into the logic of this fault tree. Each boundary condition set contains one or more BEs or house events with their associated logic states — true or false — to account for the impact of the MU condition. Changes in the logic state of the events located in the fault tree structures can modify the fault tree logic. Therefore, the same fault tree logic can be used both for SU risk evaluation and MU risk evaluation.

During the quantification of the model, each boundary condition set can be activated by three means:

(1) By IE, which will activate the boundary condition set when the corresponding event tree is being quantified;
(2) By the event tree branch node, where the boundary condition set can be assigned to the true or false branch of the event tree and activated during the quantification of this event tree based on what has occurred earlier in the sequence;
(3) Within each quantification specification defined in the RiskSpectrum code.

Using this approach, SU fault trees can be used for MU risk evaluation.

II.2.4. Human interactions

The SU HFEs were applied in the MU model. As with the other events in the model, unit specific designations were given to each HFE unless the HFE was clearly identified as a MU HFE (i.e. the action, when performed, affected multiple units, such as the decision to deploy portable equipment). The dependency between HFEs both within each unit and across units was implemented using minimal cut set (MCS) post-processing within the 'MCS case' quantification in the RiskSpectrum code. The approach to assigning dependence between HFEs across units is discussed in the Phase I MUPSA methodology document. HFE combinations that occurred in the cut sets were identified and the dependency adjustment applied as needed.

The following are the main groups of HFEs analysed for MU effects and dependency in the MUPSA modelling:

(a) Operator fails to initiate aggressive depressurization;
(b) Operator fails to initiate emergency boration;
(c) Operator fails to initiate bleed and feed;
(d) Operator fails to drain the bubble condenser following small LOCA;
(e) Operator fails to isolate small LOCA;
(f) Operator fails to perform manual scram;
(g) Operator fails to recover feedwater supply;
(h) Operator fails to restore off-site power;
(i) Plant personnel fail to recover feedwater using a mobile source.

II.2.5. Common cause analysis

In the original PSA model, the alpha factor method was used to quantify the CCF BE probabilities, based on NUREG/CR-5801 [64]. The original CCFs for one unit were used for twin unit conditions. One set of CCF BEs has been used for Units 1 and 2 (dominant combination of four components) and another set of CCF BEs has been used for Units 3 and 4 (dominant combination of six components) to account for the assumption that the components in each pair were acquired many years apart and thus are not sufficiently similar to have a significant common cause potential.

Only one CCF BE has been introduced for all four units, which represents the CCF of the mobile sources for Units 1 and 2 and Units 3 and 4, since it is assumed that these components were procured at the same time. If a more detailed MUPSA is performed, a study of the proper CCF coupling factor analysis needs to be performed, since the mobile source CCF is risk significant in the MUPSA results.

II.2.6. Seismic fragility analysis

Seismic failures of equipment and structures, including seismically induced IEs and mitigating system failures, are modelled by different BEs within the different earthquake acceleration intervals. The probabilities of these seismic failures are determined by fragility analyses. Each fragility analysis quantifies the likelihood that a component or structure may fail, as a function of the earthquake peak ground acceleration. For the case study, two sets of seismic fragilities were used — one set for the SSCs of the old units (Units 1 and 2) and the second set for the SSCs of the new units (Units 3 and 4).

Based on the discussion in the Phase I MUPSA methodology, it is assumed that both twin unit pairs (Units 1 and 2 and Units 3 and 4) are fully seismically correlated. Therefore, the fragility probabilities set from the original SUPSA model was used for twin unit conditions. One set of fragility BEs has been used for Units 3 and 4.

There are three BEs representing full seismic correlation of all four units:

(1) Seismically induced failure of LOOP due to loss of grid;
(2) Off-site power recovery;
(3) Seismically induced failure of mobile sources.

II.2.7. Model quantification

The METs developed for each IE discussed in Section II.2.1 were developed and quantified using RiskSpectrum, including the application of boundary conditions, HEP dependencies and IE specific probabilities such as fragility events. The quantified cut sets were reviewed by the steering group for the case study and the model was adjusted as needed to ensure the results were reasonable. For the case study, the quantification of each IE is performed separately, and there were no merged results other than a single merged set of cut sets for seismic events. The results of the quantification are provided in Section II.3.

II.2.8. MU ratio calculation

The MU ratios in the case study were calculated using Eq. (3):

$$R_2(\text{old}) = \frac{\text{CDF}_{12}}{\text{CDF}_1}; R_2(\text{new}) = \frac{\text{CDF}_{34}}{\text{CDF}_3}; R_4(\text{old}) = \frac{\text{CDF}_{1234}}{\text{CDF}_1};$$
$$R_4(\text{new}) = \frac{\text{CDF}_{1234}}{\text{CDF}_3} \tag{3}$$

where CDF_1 is the SUCDF of Unit 1, CDF_3 is the SUCDF of Unit 3, CDF_{12} is the MUCDF of Unit 1 and Unit 2, CDF_{34} is the MUCDF of Unit 3 and Unit 4 and CDF_{1234} is the MUCDF of all four units.

II.3. RESULTS OF THE PHASE II CASE STUDY

The results for the selected IEs analysed in the MUPSA Phase II case study are summarized in Table 3. Beside representation of SUCDF and MUCDF, Table 3 contains the results of various MU ratio calculations. These values are presented to illustrate the ratios between SUCDFs and MUCDFs and have been used for interpretation of the results.

II.3.1. Baseline MUPSA results

Table 3 provides the results of the case study quantification. The MU ratio (R) in the table was calculated based on Eq. (3) provided in Section II.2.8.

II.3.2. Sensitivity analysis

The following cases were analysed using sensitivity analysis:

(a) Radioactive release from one of the units;
(b) CCFs;
(c) Human interactions;
(d) Administrative shutdowns;
(e) Seismic correlations.

The objective of the sensitivity analysis was to check the plausible alternative assumptions stated in section 3.1 of Ref. [54] and the uncertainties connected with specific modelling approaches, and to analyse the potential level of impact from radioactive release from one unit to another. The five areas of sensitivity analysed for the case study are discussed in the following subsections.

II.3.2.1.   Radioactive release from one of the units

One of the elements of a MU risk assessment is the concept that multiple units in (or progressing towards) a core damage scenario can impact on operator actions at all units. To investigate this issue, this case study initially designed a scenario whereby one unit (i.e. Unit 1) is assumed to reach core damage before

TABLE 3. SUMMARY OF THE RESULTS OF THE PHASE II CASE STUDY

		Initiating event				
		SLBO	Fire in the turbine hall	LOOP (SFT approach)	LOOP (MET approach)	Seismic events
CDF for Units 1 and 2 (old units)	Unit 1	2.56E-08	7.65E-07	1.13E-06	1.13E-06	1.58E-04
	Unit 2	9.84E-08	2.98E-06	1.13E-06	1.13E-06	1.58E-04
	Units 1 and 2	1.87E-10	6.46E-09	1.68E-08	1.68E-08	1.32E-04
	R_2 (old)	7.30E-03	8.44E-03	1.49E-02	1.49E-02	8.35E-01
CDF for Units 3 and 4 (new units)	Unit 3	n.a.[a]	n.a.	7.48E-07	7.48E-07	2.34E-05
	Unit 4	n.a.	n.a.	7.48E-07	7.48E-07	2.34E-05
	Units 3 and 4	n.a.	n.a.	3.42E-09	3.42E-09	1.59E-05
	R_2 (new)	n.a.	n.a.	4.57E-03	4.57E-03	6.79E-01
CDF for all units	Units 1, 2, 3 and 4	n.a.	n.a.	2.45E-15	2.45E-15	1.91E-05[b]
	R_4 (old)	n.a.	n.a.	2.17E-09	2.17E-09	1.21E-01
	R_4 (new)	n.a.	n.a.	3.28E-09	3.28E-09	8.16E-01

[a] n.a.: not applicable.
[b] It is typically expected that CDF_{1234} is lower than CDF_{34}, which is not the case for the presented seismic event results. In this case, the difference between the CDF_{1234} and CDF_{34} results is conditioned by the fact that the number of MCSs for CDF_{1234} is much higher than that corresponding to CDF_{34}.

TABLE 4. RESULTS OF THE SENSITIVITY ANALYSIS RELATED TO THE CONSIDERATION OF RADIOACTIVE RELEASE FROM UNIT 1 IN CASE OF LOOP

IE	Base case	Sensitivity case HEP = 0.1	Sensitivity case HEP = 1.0
CD_{1234} for LOOP at four units	2.45E-15	2.68E-15	6.67E-14
Units 1–4 MUCDF ratio	2.17E-09	2.37E-09	5.90E-08

the other units and to have an early release from the containment while the other units are not yet in a core damage condition.

The sensitivity performed involved a four-unit LOOP, which had a relatively low MU ratio, as shown in Table 3 above. The two sensitivity runs assumed a CDF event in Unit 1 and set the longer term actions to conservatively high values — either 0.1 or 1.0 — for Units 2, 3 and 4. The results of the sensitivity cases are shown in Table 4.

In this case, the MUCDF for four units increased significantly for the scenario where the HEPs were set to 1.0. However, the MUCDF ratio is still relatively small. In reviewing the results, it was concluded that the impact of radioactive release can play a significant role in scenarios where the results are not dominated by CCF or correlated hazards. However, for the case study, all analysed IEs are dominated by CCF and correlated hazards, so the radioactive release sensitivity/assumptions does/do not greatly impact the results.

II.3.2.2.   CCFs

In the base case study, the following assumption was used to model CCF (simplistically) [5]:

"The CCF model for inter-unit CCF can be limited to "n-of-n" failures, where "n" is defined by the total number of components across all of the units affected by the specific initiating event being evaluated."

During the development of this sensitivity case, it was determined that the simplified CCF modelling for the case study may not be conservative. The n-of-n treatment for CCF did not include all CCF combinations bound by n-of-n failures. As discussed in Section 4.4.5.2, in the two-unit, four-train

TABLE 5. RESULTS OF THE SENSITIVITY ANALYSIS RELATED TO
THE CCFs

IE	Conservative	Detailed
CD_{12} at SLBO + LMF	8.05E-10	2.35E-10
CD_{1234} for LOOP at four units	4.60E-12	8.38E-14

example, if a 4-of-4 CCF event is included, the probability also needs to include the 3-of-4 combinations for the initial simplified/conservative modelling. This approach was not used for the case study, which resulted in enhancements in the methodology in this area.

The two sensitivity studies for CCF were related to the top six CCFs of the model in terms of two IEs. The sensitivity included modelling the CCF conservatively and expanding the CCF using detailed alpha factor modelling. This was performed for a two-unit scenario involving SLBO with LMF and a four-unit LOOP. The results of the sensitivity runs are provided in Table 5.

As can be seen from the results of these sensitivity cases, the selection of parameters for CCFs has a significant impact on the results of the considered IEs and hence may play a significant role in MU risk evaluation. For internal hazards, the accurate modelling of CCF parameter or dominant CCF events can significantly decrease the MU risk. It would not be expected to be the case for external hazards (e.g. seismic) due to the fact that the results may be dominated by correlations or high failure probabilities for high severity hazard events. In that case, if the MU risk is dominated by external hazards, then the conservative CCF approach is likely to be acceptable.

II.3.2.3.   Human interaction sensitivity

Specific sequences may need to be addressed where the potential exists for a time difference between core damage sequences at different units, which would potentially result in the contamination of areas needed to perform operator actions attempting to prevent core damage in Units 2 to 4. This situation considers a potential increase in HEPs for the above mentioned accident sequences. The following three sensitivity cases have been analysed:

— Case A: Increase HEP for local action in Unit 2 when Unit 1 is in a core damage sequence for SLBO with LMF IE;

TABLE 6. RESULTS OF THE SENSITIVITY ANALYSIS RELATED TO
HUMAN INTERACTIONS

IE	Base case	Sensitivity case A (HEP = 0.5)	Sensitivity case B (HEP = 0.5)	Sensitivity case C (HEP = 1.0)
CD_{12} for SLBO + LMF IE	1.87E-10	6.67E-10	—[a]	4.01E-08
CD_{12} for SLBO + LMF IE Seq. 22	1.47E-12	—	6.68E-12	—
CD_{12} for fire IE	6.46E-09	n.a.[b]	n.a.	2.34E-07
CD_{1234} for LOOP IE at four units	2.45E-15	n.a.	n.a.	8.94E-14

[a] — : data not available.

[b] n.a.: not applicable.

— Case B: Increase HEP for local action in Unit 2 only when Unit 1 is on an anticipated transient without a scram core damage sequence, and Unit 2 is not, for SLBO with LMF IE;
— Case C: Set the HEPs of all local actions equal to 1.0 and recalculate all internal IEs.

The results of these sensitivity cases are presented in Table 6.

In all three sensitivity cases, the increased HEPs lead to increased CDF of two or four units. As can be seen from the results of these sensitivity cases, local human actions have a significant impact on the results of the considered IEs and hence may play a significant role in MU risk evaluation. For internal hazards, the accurate modelling of human interactions can significantly decrease the MU risk. This may not be the case for external hazards (e.g. seismic) due to the fact that the results may be dominated by correlations or high failure probabilities for high severity hazard events.

II.3.2.4.   *Administrative shutdowns*

An assumption in the methodology currently says [5]:

"Administrative shutdowns of otherwise unaffected units do not need to be explicitly included as an initiating event in the MUPSA, including the cases when [the] affected unit approached core damage."

Therefore, this sensitivity case is intended to check the impact of this assumption by analysing the effect of the modelling of administrative shutdowns at Units 3 and 4 in the case of a fire event at Units 1 and 2 (even though Units 3 and 4 are not directly affected by the fire).

The results of this sensitivity case are presented in Table 7.

As can be seen from the results, the MUCDF for two or more core damages is negligibly affected by consideration of administrative shutdowns in Units 3 and 4 that are not affected by the IEs.

Another example of sensitivity analysis regarding administrative shutdowns was the consideration of a SLBO at Unit 1 with successful steam isolation. In this scenario, Unit 1 is in the transient conditioned by a steam line break. Meanwhile,

TABLE 7. CORE DAMAGE RESULTS (1/SITE-YEAR) OF THE SENSITIVITY ANALYSIS RELATED TO THE CONSIDERATION OF ADMINISTRATIVE SHUTDOWNS

IE	Unit 1	Unit 2	Units 1 and 2	Units 1, 2, 3 and 4	$R_{4/2}$ (old)[a]	MUCDF two or more
Fire in the turbine hall of Units 1 and 2 (no effect on Units 3 and 4)	7.65E-07	2.98E-06	6.46E-09	n.a.[b]	n.a.	6.46E-09
Fire in the turbine hall of Units 1 and 2 (administrative shutdown at Units 3 and 4)	7.65E-07	2.98E-06	6.46E-09	1.59E-14	2.45E-06	6.46E-09

[a]   This MU ratio was calculated using R4/2 (old) = CDF_{1234}/ CDF_{12}, where CDF_{12} is the MUCDF of Unit 1 and Unit 2 and CDF_{1234} is the MUCDF of all four units.

[b]   n.a.: not applicable.

Unit 2 is considered to be in the process of administrative shutdown, though it is not directly affected by the IE. A specific event tree was constructed for this sensitivity case.

The results of the sensitivity case are presented below:

— CDF for Unit 1 in the case of SLBO at Unit 1 with successful steam isolation (i.e. no effect on Units 2, 3 and 4) equals 1.59×10^{-9}/year;
— MUCDF for Unit 1 and Unit 2 in the case of SLBO at Unit 1 with successful steam isolation and administrative shutdown at Unit 2 (i.e. no effect on Units 3 and 4) equals 3.2×10^{-14}/year.

As can be seen from the results, the MUCDF for two or more cores is negligibly affected by consideration of administrative shutdowns in Unit 2 that are not directly affected by the IEs.

II.3.2.5.   Seismic correlations

In the base MUPSA model, full seismic correlation is assumed for the units within the pairs. Unit 1 is fully correlated with Unit 2, and Unit 3 is fully correlated with Unit 4. Within this sensitivity case, zero correlations for Unit 1 and Unit 2 are set, and the impact on CD_{12} for seismically induced IEs at Units 1 and 2 is recalculated.

The results of this sensitivity case are presented in Table 8.

As can be seen from the results, there is no large difference between the CDF for fully correlated Units 1 and 2 and the CDF for zero correlation of both units. This is caused by the fact that the dominant MCSs represent the failure of primary to secondary side heat removal and the failure of primary bleed and feed at both units. As mobile source failure is common for both units, for the sensitivity case new seismically induced failure of the feed-and-bleed function at Unit 2 is added to each dominant MCS. Because the probabilities of these non-correlated failures for dominant seismic intervals are quite high (about 0.7), the total contribution is not as significant as initially expected.

TABLE 8. RESULTS OF THE SENSITIVITY ANALYSIS RELATED TO THE SEISMIC CORRELATIONS

IE	Base case	Sensitivity case
CD_{12} for seismic events	1.32E-04	9.65E-05

Note that the seismic examples presented in Safety Reports Series No. 96 [1] came to a similar conclusion: assumptions regarding seismic correlation do not appreciably impact the results because of the dominant contributions from seismic events with high intensity that produce large contributions from independent seismic failures near the top of the fragility curve.

II.4.   CONCLUSIONS FOR THE PHASE II CASE STUDY

The Phase II case study described above has been implemented based on the MUPSA methodology developed within Phase I of the MUPSA project.

The case study allowed the NSNI MUPSA working group to verify various aspects of the proposed MUPSA methodology by applying it to a realistic NPP configuration using a realistic PSA model. The integrated MUPSA model has been developed within the case study, which provides an opportunity to estimate selected MU risk metrics.

With respect to specific results, the case study concluded the following, in line with the results of the ABWR pilot evaluation in the UK:

(a)   The two-unit MU ratio for non-seismic events is relatively low (~1%), as a result of the possibility of using mobile equipment to mitigate the accident and the moderate correlation/dependency between the units (e.g. CCFs, human interaction dependencies).

(b)   The four-unit MU ratio is generally very low because of the absence of correlations between old and new units.

(c)   The summary results for seismic events are presented in Fig. 9. As can be observed from the figure, the contribution of different seismic intervals changes from the old units to the new units (intervals 3, 4 and 5 are dominant for the old units, and intervals 4, 5 and 6 are dominant for the new units).

(d)   The MU ratio for seismic events is much higher than for non-seismic events as a result of a combination of effects: (a) seismic correlation between units in each pair dominates the results at the lower accelerations, and (b) high independent failure probabilities dominate the results at the higher accelerations. Overall, the difference between conservative treatment of correlation and ignoring correlation is relatively small compared with other MUPSA factors for seismic hazards.

(e)   Dependencies between operator actions and CCFs remain dominant in the results for MU. Additionally, these correlations between the units demonstrate that over 90% of MUCDF involve similar pairs of units (e.g. Units 1 and 2) in the same accident sequences.

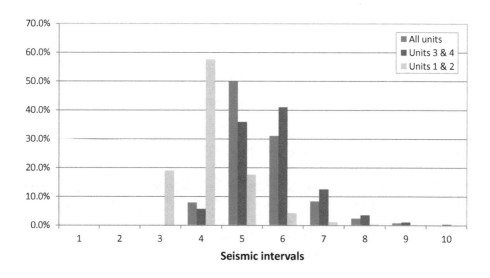

FIG. 9. The importance of individual seismic intervals for MUCDF of different combinations of units.

It could be concluded that the scope of the MUPSA methodology was sufficiently detailed at the implementation level to support the development of a case study. The methodology was applied for a four-unit NPP site with two pairs of very similar units that only differed in that the new pair of units was assumed to have some additional safety features and to be more seismically rugged. However, no inherent limitations were identified that would prevent the application of the methodology to a larger number of units or a diverse population of unit types on a site.

The case study provided a better understanding of the level of effort necessary to expand current PSA studies to the MU context and identified the main challenges for MUPSA. Looking through MU risk profiles, it could be concluded that CCFs, human interactions (HFE dependency) and seismic correlation factors are dominant in MUCDF. The sensitivity analyses illustrated that alternative assumptions in the areas of interunit CCF and HFE dependency could have a significant impact on the results.

The case study has demonstrated that the general principles of PSA can be applied in the MU context. Thus, realistic modelling is important in order to obtain a sufficient basis for risk informed decision making. In this context, it is necessary to highlight that both positive and negative interactions among units have been considered, which allowed analysts to generate a realistic risk profile as an output of the case study.

The case study provided feedback on the applicability of the proposed methodology for standard PSA tasks. These recommendations are discussed in Section 2.4.2 of this safety report. Reviewing the results obtained for LOOP IE, it can be concluded that both approaches for combining the accident sequences (the MET approach and the single-top fault tree approach) are applicable for the construction of MU event trees and lead to the same results.

The IAEA/NSNI MUPSA working group members concluded that the case study has been successfully implemented and served its purpose. It completes the efforts implemented in the project's Phase I and supports the finalization of the project in Phase III (improvement of the methodology based on the feedback from the case study and publication of this safety report).

II.5. ENHANCEMENTS TO THE MUPSA METHODOLOGY BASED ON INSIGHTS FROM THE PHASE II CASE STUDY

As an outcome of the MUPSA Phase II case study discussed in Section 2.4.2, a number of recommendations for changes to the MUPSA methodology were provided. The recommended changes can be summarized as follows:

(1) Elaborate more on the principles considering the impact of radioactive release from one unit on human actions in other units and on sensitivity studies related to this topic (see Sections 4.4.4 and 4.6.3).
(2) Elaborate more on the assumption related to the consequential LOOP, in particular what needs to be done in the MU context if the probability of consequential LOOP is high at a particular site (see Sections 4.3.1.1 and 4.4.3).
(3) Describe the screening principles for POSs and the consideration of the unit availability factor in IE frequency assessment (see Section 4.2.1.2).
(4) Present available experience on partial hazard correlations in the MU context (see Section 4.4.5.3).
(5) Elaborate more on expanding the CCF groups from SU to MU. Particularly describe the fact that application of the 'n out of n' CCF model needs to be done in such a way as to ensure that CCF probabilities are conservatively calculated (see Section 4.4.5.2).
(6) Describe the principles on the selection of scenarios where the impact of radioactive release from one unit to another needs to be considered (e.g. systematic search of relevant sequences considering the dynamics and timing issues) (see Section 4.4).
(7) Revise Appendix I based on lessons learned (level of MUPSA efforts) (see Appendix I).

(8) Describe in more detail normalizing the IE frequencies for the MU context (e.g. for IEs that could be either SU or MU, depending on the specific severity, how to preserve the portion of the original IE frequency related to MUIE) (see Section 4.3.4).

(9) Elaborate on SUIEs that could potentially be MUIEs in the case of CCF (see Section 4.3.1).

(10) Discuss the need to revisit screening criteria related to the 0.1% of SUCDF for cases when one of the hazards is dominating (see Section 4.2.1.3).

(11) Supplement and elaborate more detail in section 3.3 (analysis and interpretation of the risk profile of a SUPSA) of Ref. [54] with feedback from the preparation of the input deck for the case study.

(12) Elaborate on how the damage context may be different at each unit and how this needs to be considered in adjusting HEPs and dependency levels. For example, external hazards with different levels of magnitude (e.g. different earthquake levels) might affect individual units in differing ways, which can lead to varying implications for human performance.

The above recommendations are incorporated into the methodology provided in Section 4. For some of the items described above, there is a lack of experience and approaches, and therefore they need more investigation (e.g. HRA, CCF). These items have been grouped by topic and are briefly described in Section 6, which provides considerations on the path forward for MUPSA.

Apendix III

ACCIDENT SEQUENCE MODELLING APPROACHES

As noted in the IAEA case study, there is no single approach to modelling MUPSA accident sequences. During the Phase I and Phase II development, the accident sequence modelling used either in the ABWR study or the IAEA case study included three general approaches:

(a) MET approach: This approach develops a single event tree for each modelled IE, with the event tree tops using the same or similar tops as used in the SUPSA. As with a SUPSA, due to the size of the event trees, accident sequence transfers may result in multiple event trees for each IE.
(b) SFT approach: This approach develops a SFT model for each risk metric, which combines a SFT model for each of the SUPSA models.
(c) Hybrid MET/SFT approach: This approach develops as single event tree for each modelled IE, but the event tree tops use single-top fault trees for each unit modelled in the MUPSA.

Note that the methodology referred to in the case study as a SFT approach, used for the four-unit LOOP analysis, is referred to above as the hybrid approach. The following sections provide an overview of these approaches.

III.1. MET MODELLING APPROACH

The objective of the MET approach is to model the plant response (e.g. technological equipment, operator actions) for each IE in the MUPSA using a single event tree using the existing SUPSA top events. In the MU context, the accident sequences are assumed to be already available from the single reactor PSAs, and the method is expected to address different possible responses to IEs involving two or more reactor units.

The application of the MET approach includes the following steps:

Step 1: Identification of the sequences leading to the undesired event (e.g. core damage) for each unit for corresponding IE

As described in Section 4.2.3, the MUIE could trigger similar as well as different IEs for different units on-site. In accordance with the underlying assumption, it is assumed that the accident sequence models (e.g. event trees) are available for each SU and address the plant response to triggered IEs. At this step, those accident analysis models need to be thoroughly investigated in order to identify the sequences leading to the undesired event (e.g. core damage). This investigation allows the analyst to simplify the accident sequences; in other words, to identify the accident sequences that could be potentially neglected because of the negligible risk contribution (see the screening discussion in Section 4.4.1).

Step 2: Conversion of the accident sequence models for individual units into the event tree logic

Once the accident sequences have been identified and understood, the SU accident sequence models need to be converted into the event tree logic. The accident sequence development would depend on what risk metrics are selected for the MUPSA. For example, for a two-unit site, if the MET is developed to include site CDF, including the analysis of SUCDF sequences, then the Unit 2 core damage sequences would be developed for both core damage and non-core damage sequences from the Unit 1 event tree.

An example of MET for four units is presented in Fig. 10. The example was developed from the IAEA case study, which has two sets of identical units, including two older design units and two newer design units. The MET represents a two-unit LOOP on Units 1 and 2. In this example, the Unit 2 accident sequences are not added to the non-core damage accident sequences for Unit 1, since the risk metric of concern is MUCDF.

The example in Fig. 10 includes all four possible MUCDF sequences. The MET approach can provide detailed calculations for all combinations and potentially provide insights into the relative level of dependence (or independence) between units based on the resulting accident sequence frequencies and cut sets.

Step 1 above requires that each IE with potential MU impacts be modelled using a single MET. In the example case above, this would include the same event (LOOP) occurring on other units or resulting in a degraded condition on the second unit.

As identified in the case study, the MET developed above would not model all permutations/combinations for core damage. In this case, the non-MU sequences calculate the frequency of Unit 1 core damage with Unit 2 not going to core damage. The sequence for Unit 2 in core damage and Unit 1 not in core damage would require adding the Unit 2 logic to all success branches from the Unit 1 event tree. Additionally, the logic starts with a Unit 1 core damage event and there may be some difference if Unit 2 is the unit initially in core damage. The methodology in Section 4 discusses methods to account for this limitation.

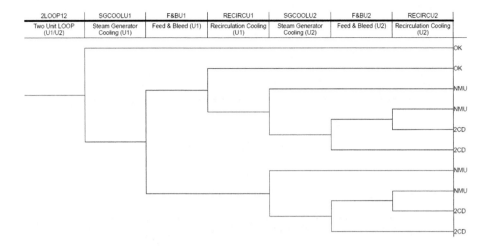

2LOOP12	SGCOOLU1	F&BU1	RECIRCU1	SGCOOLU2	F&BU2	RECIRCU2
Two Unit LOOP (U1/U2)	Steam Generator Cooling (U1)	Feed & Bleed (U1)	Recirculation Cooling (U1)	Steam Generator Cooling (U2)	Feed & Bleed (U2)	Recirculation Cooling (U2)

FIG. 10. The example of MET for two units for a MU LOOP (OK – success state, NMU – non-MUCD sequences).

III.2. SFT MODELLING APPROACH

The objective of the SFT modelling approach is to model the MUPSA accident sequence model using a single-top event for each hazard group. The SFT approach, also sometimes referred to as a top logic modelling approach, combines the risk logic for all units, all hazards and all modes under a single-top logic gate. If multiple risk measures are desired, then each risk measure can be modelled using a SFT top gate. This can include, for example, a single top for each combination of unit risk. In the example shown in Fig. 11, this could include (for example) either a single-top event (e.g. CDF in two or more units) or three separate top events for CDF in two, three or four units. If desired, each MUCDF sequence shown in Fig. 11 could be modelled with a SFT top event. However, to ensure the sequence frequencies are correct, some modelling correction is needed, such as removal of four-unit CDF cut sets from the three-unit cut set results. Typically, this would be modelled either by using a NOT gate or by deleting common cut sets from the three-unit cut set files.

The application of the SFT approach includes the following steps:

Step 1: Identification of the sequences leading to the undesired event (e.g. core damage) for each unit for corresponding IE

As described in Section 4.2.3, the MUIE could trigger similar as well as different accident sequences for different units on-site. This investigation

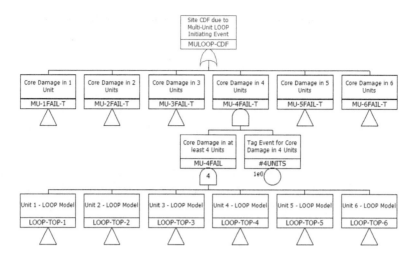

FIG. 11. SFT approach for six units.

allows the analyst to simplify accident sequences, based on the screening of accident sequences that are negligible risk contributors.

Step 2: Conversion of the accident sequence models for individual units into the fault tree logic

Once the accident sequences have been identified and understood, the accident sequence models need to be converted into a single-top logic fault tree for each unit based on an undesired event (e.g. core damage) for the relevant unit. A single-top model may already be developed for the PSA or automatically created using the PSA software once the IE scope has been determined.

Step 3: Construction of SFT for the MU site

Once the fault trees representing the core damage for each unit are available, they can be combined under a SFT top event representing the risk metric within the scope of the MUPSA. This may require multiple top events be developed, if the logic for selected hazards is analysed separately. Additionally, if there are multiple risk metrics selected, each risk metric is represented by a single-top event (for each analysed hazard group).

This method requires the development of a single-top logic model for each unit PSA scope/hazard and the creation of a single OR gate for each unit (combining the individual scopes). The methodology is to combine the logic for each unit under a single-top event gate. This top event could, for example, combine all POSs and hazards (such as at-power, shutdown, refuelling, fire, flood and seismic) for each unit under one top logic event (i.e. SCDF). The top events for each unit are then combined under logic representing each of the selected risk metrics. For example, for a two-unit site, an AND gate of each unit's top event fault tree is developed for MUCDF. For site CDF, the top event would be an OR gate of each SU top logic fault tree. From this model, the SUCDF for the site and the MUCDF when all units have core damage can be determined.

An example of a SFT model approach is provided in Fig. 11, taken from Annex IV. In this example, top events are developed for different MUCDF risk measures and the top event for each risk metric is a combination gate for the number of units. The developed logic for the four-unit CDF is shown for LOOP only, but this can be combined for each or all of the hazards, assuming the hazards can be analysed simultaneously (e.g. some hazards, such as fire or seismic, are traditionally analysed separately due to software and calculation limitations).

In the example in Fig. 11, if the risk metric necessitated a specific combination to be calculated, such as CDF in four, and only four, units, additional fault tree logic would be needed to delete the CDF from additional unit CDFs (e.g. CDF in five or six units). This step can also be done through post-processing. This step is important when calculating SUCDF, to remove double counting of the results where cut sets can cause CDF on either one or more than one unit.

Although the method is referred to as a SFT method, it requires a top event to be developed for each of the risk metrics being calculated, such as MUCDF, SULRF or MULRF. Additionally, if the site involves more than two units, separate top events for MUCDF/MULRF may be desirable to determine, for example, MUCDF involving two or more units, MUCDF involving three or more units, etc. There are not currently separate risk metrics for different combinations of units, although the case study application in Phase 2 of this effort calculated the likelihood of both two- and four-unit CDF (for a four-unit site).

Only one IE is allowed in each cut set during quantification. If the IE is a MUIE (e.g. simultaneously affecting multiple units), then the SFT model would combine the sequence logic automatically. The SFT logic would include combinations of IE specific accident sequences for multiple units and estimate the risk for both conditions when two (or more) units are experiencing the same accident sequence or when units are experiencing different accident sequences. Both the ABWR pilot evaluation in the UK and the IAEA case study showed that, once MU CCF modelling is included in MUPSA, similar units will experience the same accident sequence when a MUCDF event occurs. However, when units are dissimilar and there is little CCF predicted between units, the units may be just as likely to experience difference accident sequences as similar accident sequences. In the case of the accident at the Fukushima Daiichi NPP, the three dissimilar units 1, 2 and 3 experienced different sequences.

When the MU sequence involves a SUIE on one unit and a forced shutdown on the other unit, the SFT logic may need to be developed differently in order to have the MUPSA correctly estimate the risk. In general, the SUIEs are screened from the MUPSA unless another unit on-site is impacted, resulting in a degraded condition causing the plant(s) to be tripped. See Section 4.2.1 above for a discussion on the screening of SUIEs. If a SUIE is retained in the model, there are two general approaches that can be used for the SFT approach:

(1) CCDP approach: One approach is to develop a SFT logic for the units being shut down, which calculates a CCDP (e.g. a conditional large release probability). The CCDP logic can be based on the manual shutdown/ trip accident sequence logic, but with the trip frequency set to 1.0 (true). However, if the degraded condition can lead to another IE for the second and subsequent units, the CCDP model may include alternate accident sequence logic. The SFT logic would then include an AND gate of the SUIE logic for the first unit, followed by the CCDP for the other units that are being manually shut down.

(2) Create new logic for the SUIE affecting another unit: This approach involves the creation of new SU top logic for the accident sequences for the other units experiencing a manual shutdown, trip or degraded condition. This logic can be based on the existing SUPSA but altered as needed to account for the degraded condition. The new logic would include the SUIE in the fault tree top logic.

When creating a MUPSA using SFT logic, the MUPSA can be built in stages to test the approach prior to developing the entire MUPSA. Creating a SU top logic model with a small number of key MUIEs will give a good indication of the feasibility of the method. The SU models supporting the MU SFT model can be built by adding additional IEs, hazards and POSs one at a time. With each addition, the top logic model can be tested by quantifying. Review of the results can then be performed, resulting in model refinements as needed prior to modelling all IE and accident sequence logic. For example, the initial MUPSA model may include simplified CCF modelling, which can then be refined upon the initial quantification involving the top SU accident sequences involving a MUIE.

III.3.   HYBRID APPROACH — COMBINED MET/SFT MODELLING
       APPROACH

This approach develops as single event tree for each modelled IE, but the event tree tops use single-top fault trees for each unit modelled in the MUPSA.
The hybrid approach includes the following steps:

Step 1: Identification of the sequences leading to the undesired event (e.g. core damage) for each unit for corresponding IE

As described in Section 4.2.3, the MUIE could trigger similar as well as different accident sequences for different units on-site. This investigation allows analysts to simplify accident sequences, based on the screening of accident sequences that are negligible risk contributors.

Step 2: Conversion of the accident sequence models for individual units into the fault tree logic

Once the accident sequences have been identified and understood, the accident sequence models need to be converted into a single-top logic fault tree for each IE for each unit based on an undesired event (e.g. core damage) for the relevant unit.

Step 3: Construction of MET for the MU site

Once the fault trees representing the core damage for each unit are available, they can be combined under a MET. The accident sequence development would depend on what risk metrics are selected for the MUPSA. For example, for a two-unit site, if the MET is developed to include site CDF including the analysis of SUCDF sequences, then the Unit 2 core damage sequences would be developed for both core damage and non-core damage sequences.

This method necessitates the development of a single-top logic model for each unit PSA scope/hazard or IE. An example of a hybrid MET/SFT model approach is provided in Figs 12 and 13.
Figure 12 shows the Step 2 development of a top logic fault tree for each IE for each unit. This is similar to the top logic development approach for the SFT method, although with the SFT method, a single top is developed for all IEs (combined). Figure 13 shows an example MET using the hybrid approach for four units.

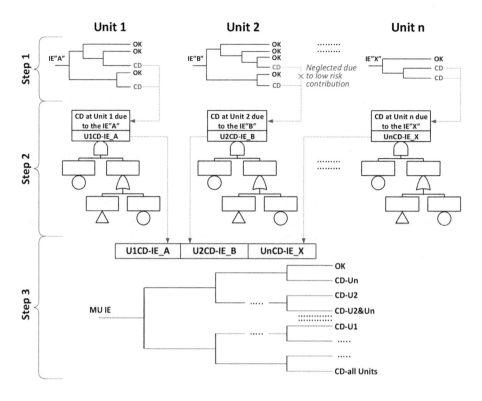

FIG. 12. Top logic development for MUPSA hybrid approach.

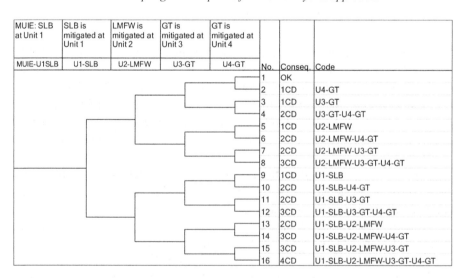

MUIE: SLB at Unit 1	SLB is mitigated at Unit 1	LMFW is mitigated at Unit 2	GT is mitigated at Unit 3	GT is mitigated at Unit 4			
MUIE-U1SLB	U1-SLB	U2-LMFW	U3-GT	U4-GT	No.	Conseq.	Code
					1	OK	
					2	1CD	U4-GT
					3	1CD	U3-GT
					4	2CD	U3-GT-U4-GT
					5	1CD	U2-LMFW
					6	2CD	U2-LMFW-U4-GT
					7	2CD	U2-LMFW-U3-GT
					8	3CD	U2-LMFW-U3-GT-U4-GT
					9	1CD	U1-SLB
					10	2CD	U1-SLB-U4-GT
					11	2CD	U1-SLB-U3-GT
					12	3CD	U1-SLB-U3-GT-U4-GT
					13	2CD	U1-SLB-U2-LMFW
					14	3CD	U1-SLB-U2-LMFW-U4-GT
					15	3CD	U1-SLB-U2-LMFW-U3-GT
					16	4CD	U1-SLB-U2-LMFW-U3-GT-U4-GT

FIG. 13. Example MET for MUPSA hybrid approach (SLB – steam line break, LMFW – loss of main feedwater, GT – general transient).

114

In this four-unit example, the end states include core damage from one, two, three and four units. With quantification, accounting for sequence success and failure logic, the end states can provide an accurate assessment of each end state. The two core damage sequences can be summed to determine the frequency for a two-unit CDF event at this example site.

III.4.   ADVANTAGES AND DISADVANTAGES OF EACH
MODELLING APPROACH

As demonstrated in the IAEA case study, each approach can have the same results. Each approach can provide a calculation for any risk metric and for any combination of units on the site. Finally, each approach requires the modification of the system models, HEPs and PSA data (CCF, correlation, etc.) in order to provide accurate results. The modification of the models is similar for both, as discussed in Section 4.4.2.

The main advantage of the MET approach is the calculation of specific combinations of MU risk, including SU risk outcomes, without additional modelling or post-processing using the existing SU top events. In the example event tree shown in Fig. 10, the outcome of the event tree included SUCDF for each of the units, CDF for combinations of two units, CDF for three units and finally CDF for all four units. The cut sets for each outcome can also be developed and reviewed. If this level of information is important, a MET can provide an easier approach to determining the outcome.

The MET approach can also more directly account for event tree branch success, which allows for the removal of MUCDF cut sets for SUCDF accident sequences, or four-unit CDF from two-unit CDF sequences, for example. This can (and needs to be) accomplished using the SFT approach but requires either post-processing or the use of NOT gates in the top logic modelling.

The MET approach also provides a representation of the MU accident sequences that can occur during a MU event. This MET is easier to communicate to non-PSA individuals than the other approaches.

Although the MET approach can also calculate the risk metrics for SU outcomes, the modifications to the PSA model to account for MU aspects may result in some inaccuracies of the resulting SUPSA estimates [28]. For example, in inserting MU CCF events into the PSA, the use of simplified CCF BEs may not greatly impact the MUPSA metrics but could result in significant impact to the base CDF for the SUPSA.

The other disadvantage of the MET approach is that a MET is developed for each MUIE within the scope of the MUPSA. This adds some effort to the modelling, as a typical PSA usually involves a dozen to several dozen event trees, depending on the scope and whether Level 2 modelling is performed.

One main advantage of the SFT approach is that the model can more quickly be developed and analysed, especially if the SUPSA is already in an SFT model. The SFT model can be developed to combine all MUIEs within the scope of the MUPSA. The SFT model can also be used to calculate any risk outcome including combinations of units, although specific top logic would need to be developed for each desired outcome. However, the calculation of a specific outcome would require specific fault tree logic to remove contributions from other outcomes. For example, if developing a SFT for CDF of three units for a four-unit site, the cut sets for four units would need to be deleted from the three-unit results.

Another main advantage of the SFT approach is that if the site involves many units, the SFT approach can more easily be developed than the MET model. In particular, when the site involves more than three units, development of the MET model may be very complicated, especially considering the number of outcomes that are possible (e.g. combinations of CDF for two or more units). For several Korean sites, which include six or more units on a given site (see Annex IV and Fig. 11), the SFT approach would be much easier to use than the MET approach. The SFT approach can also be easily applied to the Level 2 modelling, as discussed in Section 5.

The final advantage of the SFT approach would involve the combination of sources, including consideration for multiple POSs for each unit. Annex IV discusses the use of the single-top logic approach for combining multiple sources from a single site. The same approach can be used to combine POS combinations for multiple units. For example, if the MUPSA would need to consider full power and multiple POSs for shutdown, then the SFT approach can easily combine the logic under a SFT gate.

The disadvantage of the SFT model method versus the MET method is that the specific combination of unit outcomes is not as easily calculated using the SFT approach. Post-processing can develop these results, but this can be complicated. To address this problem, the SFT approach used in the Republic of Korea models all possible combinations of unit outcomes under a single-top logic gate. The fault trees under the top logic gate are distinguished by the number of failed units (e.g. core damage units), and tag events with a probability of 1.0 for each number of failed units (e.g. '#1UNIT', '#2UNITS') are used to prevent the cut sets of more units failing from being subsumed by the cut sets of fewer units failing. This modelling approach makes it possible to calculate a given risk metric for the top gate (e.g. site CDF), as well as for each number of failed units (e.g. one-, two-, three- and four-unit CDF) with a single quantification [34]. An example of this modelling approach can be found in Annex IV.

Another disadvantage of the SFT approach (as well as the hybrid approach) is that the logic is harder to understand for non-PSA reviewers.

The main advantage of the hybrid approach is that it is possible to easily develop the specific outcome from specific combinations of units. Figure 13 shows the modelling for specific combinations of units going to CDF. Similarly to the SFT approach, if the SUPSA already includes SFT models, the development of the hybrid MET is relatively simple.

The main disadvantage of the hybrid approach is the need to develop separate event trees for each IE. In theory, a single hybrid MET could be developed for the MUPSA. However, the logic may need to be modified for both the IE and SFT logic supporting a single MET.

For all approaches, it can be difficult to combine the risk from multiple hazards, since different solution methods or different solution codes may be used in the quantification of each hazard. Significant modelling refinements may be necessary to develop a SFT or MET model for either SUPSA or MUPSA.

REFERENCES

[1]   INTERNATIONAL ATOMIC ENERGY AGENCY, Technical Approach to Probabilistic Safety Assessment for Multiple Reactor Units, Safety Reports Series No. 96, IAEA, Vienna (2019).

[2]   INTERNATIONAL ATOMIC ENERGY AGENCY, Consideration of External Hazards in Probabilistic Safety Assessment for Single Unit and Multi-unit Nuclear Power Plants, Safety Reports Series No. 92, IAEA, Vienna (2018).

[3]   INTERNATIONAL ATOMIC ENERGY AGENCY, Attributes of Full Scope Level 1 Probabilistic Safety Assessment (PSA) for Applications in Nuclear Power Plants, IAEA-TECDOC-1804, IAEA, Vienna (2016).

[4]   INTERNATIONAL ATOMIC ENERGY AGENCY, Risk Aggregation for Nuclear Installations, IAEA-TECDOC-1983, IAEA, Vienna (2021).

[5]   COMAN, O., POGHOSYAN, S., "IAEA project: multiunit probabilistic safety assessment", in Proc. Probabilistic Safety Assessment and Management Conf. (PSAM14), Los Angeles, CA, 16–21 Sept. 2018.

[6]   INTERNATIONAL ATOMIC ENERGY AGENCY, Safety of Nuclear Power Plants: Design, IAEA Safety Standards Series No. SSR-2/1 (Rev. 1), IAEA, Vienna (2016).

[7]   INTERNATIONAL ATOMIC ENERGY AGENCY, Safety Assessment for Facilities and Activities, IAEA Safety Standards Series No. GSR Part 4 (Rev. 1), IAEA, Vienna (2016).

[8]   FOOD AND AGRICULTURE ORGANIZATION OF THE UNITED NATIONS, INTERNATIONAL ATOMIC ENERGY AGENCY, INTERNATIONAL CIVIL AVIATION ORGANIZATION, INTERNATIONAL LABOUR ORGANIZATION, INTERNATIONAL MARITIME ORGANIZATION, INTERPOL, OECD NUCLEAR ENERGY AGENCY, PAN AMERICAN HEALTH ORGANIZATION, PREPARATORY COMMISSION FOR THE COMPREHENSIVE NUCLEAR-TEST-BAN TREATY ORGANIZATION, UNITED NATIONS ENVIRONMENT PROGRAMME, UNITED NATIONS OFFICE FOR THE COORDINATION OF HUMANITARIAN AFFAIRS, WORLD HEALTH ORGANIZATION, WORLD METEOROLOGICAL ORGANIZATION, Preparedness and Response for a Nuclear or Radiological Emergency, IAEA Safety Standards Series No. GSR Part 7, IAEA, Vienna (2015).

[9]   INTERNATIONAL ATOMIC ENERGY AGENCY, Nuclear and Radiation Safety, Resolution GC(62)/RES/6, IAEA, Vienna (2018).

[10]   INTERNATIONAL ATOMIC ENERGY AGENCY, Measures to Strengthen International Cooperation in Nuclear, Radiation, Transport and Waste Safety, Resolution GC(61)/RES/8, IAEA, Vienna (2017).

[11]   HLAVAC, P., KOVAKS, Z., A. MAIOLI, A., HENNEKE, D., AMICO, P., BONEHAM P., COMAN, O., POGHOSYAN, S., "Development of multiunit PSA model for the case study of the IAEA project" (Proc. Int. Topical Mtg on Probabilistic Safety Assessment and Analysis (PSA 2019), Charleston, SC), American Nuclear Society, La Grange Park, IL (2019).

[12] CANADIAN NUCLEAR SAFETY COMMISSION, Summary Report of the Int. Workshop on Multi-Unit Probabilistic Safety Assessment, CNSC, Ottawa, ON (2014).

[13] NUCLEAR ENERGY AGENCY/COMMITTEE ON THE SAFETY OF NUCLEAR INSTALLATIONS, Status of Site-Level (Including Multi-Unit) Probabilistic Safety Assessment Developments, NEA/CSNI/R(2019)16, OECD/NEA/CSNI, Paris (2019).

[14] INTERNATIONAL ATOMIC ENERGY AGENCY, Human Reliability Analysis for Nuclear Installations, IAEA Safety Reports Series, Vienna (in preparation).

[15] PICKARD, LOWE AND GARRICK, Seabrook Station Probabilistic Safety Assessment, PLG-0300, PLG, Irvine, CA (1983).

[16] AMERICAN SOCIETY OF MECHANICAL ENGINEERS/AMERICAN NUCLEAR SOCIETY, Addenda to ASME/ANS RA-S–2008: Standard for Level 1/Large Early Release Frequency Probabilistic Risk Assessment for Nuclear Power Plant Applications, ASME/ANS RA-Sa–2009, ANSI/ANS, New York, NY (2009).

[17] AMERICAN SOCIETY OF MECHANICAL ENGINEERS/AMERICAN NUCLEAR SOCIETY, Probabilistic Risk Assessment Standard for Advanced Non-LWR Nuclear Power Plants, RA-S-1.4–2013, ASME/ANS, New York, NY (2013).

[18] VECCHIARELLI, J., DINNIE, K., LUXAT, J., Development of a Whole-Site PSA Methodology, COG-13-9034 R0, COG, Toronto, ON (2014).

[19] KUMAR, M., et al., Advanced Safety Assessment Methodologies: Extended PSA, Lessons of the Fukushima Daiichi accident for PSA, ASAMPSA_E/WP30/D30.2/2017-32, IRSN, Fontenay-aux-Roses, France (2017).

[20] WIELENBERG, A., et al., Advanced Safety Assessment Methodologies: Extended PSA, Risk Metrics and Measures for Extended PSA, ASAMPSA_E/WP30/D30.7/2017-31 Vol. 3, IRSN, Fontenay-aux-Roses, France (2017).

[21] WIELENBERG, A., et al., Advanced Safety Assessment Methodologies: Extended PSA, Methodology for Selecting Initiating Events and Hazards for Consideration in an Extended PSA, ASAMPSA_E/WP30/D30.7/2017-31 Vol. 2, IRSN, Fontenay-aux-Roses, France (2017).

[22] KAHIA, S., et al., Advanced Safety Assessment Methodologies: Extended PSA, Report 6, Guidance Document: Man-Made Hazards and Accidental Aircraft Crash Hazards Modelling and Implementation in Extended PSA, ASAMPSA_E/WP21&WP22/D50.20/ 2017-38, IRSN, Fontenay-aux-Roses, France (2017).

[23] CAZZOLI. E., et al., Advanced Safety Assessment Methodologies: Extended PSA, Implementing External Events Modelling in Level 2 PSA, ASAMPSA_E/WP40/D40.7/2017-39 Vol. 2, IRSN, Fontenay-aux-Roses, France (2017).

[24] ELECTRIC POWER RESEARCH INSTITUTE, Aggregation of Quantitative Risk Assessment Results — Comparing and Manipulating Risk Metrics, EPRI 1010068, EPRI, Palo Alto, CA (2005).

[25] ELECTRIC POWER RESEARCH INSTITUTE, An Approach to Risk Aggregation for Risk-Informed Decision-Making, EPRI 3002003116, EPRI, Palo Alto, CA (2015).

[26] FERRANTE, F., "Consideration of multi-unit risk aspects within an integrated risk-informed decision-making framework" (Proc. Int. Topical Mtg on Probabilistic Safety Assessment and Analysis (PSA 2019), Charleston, SC, 2019), American Nuclear Society, La Grange Park, IL (2019).

[27] ELECTRIC POWER RESEARCH INSTITUTE, Framework for Assessing Multi-Unit Risk to Support Risk-Informed Decision-Making: Phase 1: Initial Framework Development, EPRI 3002015991, EPRI, Palo Alto, CA (2019).

[28] HIROKAWA, N., HENNEKE, D., "Multi-unit probabilistic safety analysis for Wylfa Newydd site", in Proc. OECD/NEA Int. Workshop on Status of Site Level PSA (including Multi-Unit PSA) Developments, Munich, 18–20 Jul. 2018.

[29] ORGANISATION FOR ECONOMIC CO-OPERATION AND DEVELOPMENT/NUCLEAR ENERGY AGENCY, ICDE Topical Report: Collection and Analysis of Multi-Unit Common-Cause Failure Events, NEA/CSNI/R(2019)6, OECD/NEA, Paris (2019).

[30] LIM, H.G., KIM, D.S., HAN, S.H., Development of logical structure for multi-unit probabilistic safety assessment, Nucl. Eng. Technol. **50** (2018) 1210–1216.

[31] KIM, D.S., HAN, S.H., PARK, J.H., LIM, H.G., KIM, J.H., Multi-unit Level 1 probabilistic safety assessment: approaches and their application to a six-unit nuclear power plant site, Nucl. Eng. Technol. **50** (2018) 1217–1233.

[32] CHO, J., HAN, S.H., KIM, D.S., LIM, H.G., Multi-unit Level 2 probabilistic safety assessment: approaches and their application to a six-unit nuclear power plant site, Nucl. Eng. Technol. **50** (2018) 1234–1245.

[33] KIM, S.Y., HAN, Y.H., KIM, D.S., LIM, H., Multi-unit Level 3 probabilistic safety assessment: approaches and their application to a six-unit nuclear power plant site, Nucl. Eng. Technol. **50** (2018) 1246–1254.

[34] HAN, S.H., OH, K., LIM, H.G., YANG, J.E., AIMS-MUPSA software package for multi-unit PSA, Nucl. Eng. Technol. **50** (2018) 1255–1265.

[35] KIM, D., PARK, J.H., LIM, H.G., A pragmatic approach to modelling common cause failures in multi-unit PSA for nuclear power plant sites with a large number of units, Reliab. Eng. Syst. Saf. **195** (2020) 106739.

[36] KIM, D.S., PARK, J.H., LIM, H.G., "Approach to inter-unit common cause failure modelling for multi-unit PSA", in Proc. OECD/NEA Int. Workshop on Status of Site Level PSA (including Multi-Unit PSA) Developments, Munich, 18–20 Jul. 2018.

[37] JEON, H., OH, K., PARK, J., "Preliminary modelling approach for multi-unit PSA in Korea (utility side)", in Proc. OECD/NEA Int. Workshop on Status of Site Level PSA (including Multi-Unit PSA) Developments, Munich, 18–20 Jul. 2018.

[38] JEON, H., OH, K., BAHNG, K., "Evaluation plan for multi-unit PSA considering operating modes in a reference site", in Trans. Korean Nuclear Society Spring Mtg, Jeju, Korea, 17–19 May 2017.

[39] OAK RIDGE NATIONAL LABORATORY, Initiating Events for Multi-Reactor Plant Sites, ORNL/TM-2014/533, ORNL, Oak Ridge, TN (2014).

[40] OFFICE OF NUCLEAR REGULATION, Safety Assessment Principles for Nuclear Facilities, Revision 0, ONR, Bootle, UK (2014).

[41] OFFICE OF NUCLEAR REGULATION, Research Report on Multi-Unit Level 3 PSA — Executive Summary, JA-ONR-1901-EX, ONR, Bootle, UK (2019), http://www.onr.org.uk/documents/2019/onr-rrr-084.pdf

[42] DROUIN, M., Technical Analysis Approach Plan for Level 3 PRA Project. Rev. 0B — Working Draft, US NRC, Washington, DC (2013).

[43] KIM, D.S., PARK, J.H., LIM, H.G., "A method for considering numerous combinations of plant operational states in multi-unit probabilistic safety assessment" (Proc. Int. Topical Mtg on Probabilistic Safety Assessment and Analysis (PSA 2019), Charleston, SC, 2019), American Nuclear Society, La Grange Park, IL (2019).

[44] SCHROER, S., MODARRES, M., An event classification schema for evaluating site risk in a multi-unit nuclear power plant probabilistic risk assessment, Reliab. Eng. Syst. Saf. **117** (2013) 40–51.

[45] LE DUY, T.D., VASSEUR, D., SERDET, E., "Multi units probabilistic safety assessment: methodological elements suggested by EDF R&D", in Proc. 12th Int. Probabilistic Safety Assessment and Management Conf. (PSAM12), Honolulu, HI, 22–27 Jun. 2014.

[46] STUTZKE, M., "Scoping estimates of multiunit accident risk", in Proc. 12th Int. Probabilistic Safety Assessment and Management Conf. (PSAM12) Honolulu, HI, 22–27 Jun. 2014.

[47] KIM, D.S., HAN, S.H., PARK, J.H., LIM, H.G., KIM, J.H., Multi-unit Level 1 probabilistic safety assessment: approaches and their application to a six-unit nuclear power plant site, Nucl. Eng. Technol. 50 (2018) 1217–1233.

[48] INTERNATIONAL ATOMIC ENERGY AGENCY, Hierarchical Structure of Safety Goals for Nuclear Installations, IAEA-TECDOC-1874, IAEA, Vienna (2019).

[49] INTERNATIONAL ATOMIC ENERGY AGENCY, Development and Application of Level 1 Probabilistic Safety Assessment for Nuclear Power Plants, IAEA Safety Standards Series No. SSG-3, IAEA, Vienna (2010).

[50] INTERNATIONAL ATOMIC ENERGY AGENCY, Development and Application of Level 2 Probabilistic Safety Assessment for Nuclear Power Plants, IAEA Safety Standards Series No. SSG-4, IAEA, Vienna (2010).

[51] INTERNATIONAL ATOMIC ENERGY AGENCY, Evaluation of Seismic Safety for Existing Nuclear Installations, IAEA Safety Standards Series No. NS-G-2.13, IAEA, Vienna (2009).

[52] HENNEKE, D., "Simplified methodology for multi-unit probabilistic safety assessment (PSA) modelling" (Proc. Int. Topical Mtg on Probabilistic Safety Assessment and Analysis (PSA 2019), Charleston, SC, 2019), American Nuclear Society, La Grange Park, IL (2019).

[53] SEUNGWOO, L., RYUM, A., CHO, N., KIM, S., KIM, H., KIM, D., "A study for identifying multi-unit initiating event and estimating frequency", in Proc. 14th Int. Probabilistic Safety Assessment and Management Conf. (PSAM14), Los Angeles, CA, 16–21 Sept. 2018.

[54] INTERNATIONAL ATOMIC ENERGY AGENCY, Working Material, Methodology for Multiunit Probabilistic Safety Assessment, Division of Nuclear Installation Safety (NSNI) Project on Multiunit PSA Phase I, IAEA, Vienna (2018).

[55] JUNG, W.S., YANG, J.E., HA, J., A new method to evaluate alternate AC power source effects, Reliab. Eng. Syst. Saf. **82** (2003) 165–172.

[56] MODARRES, M., ZHOU, T., MASSOUD, M., Advances in multiunit nuclear power plant probabilistic risk assessment, Reliab. Eng. Syst. Saf. **15** (2017) 87–100.

[57] BRUNETT, A.J., Success Criteria and Passive System Reliability Analysis, ANL-GEH-004, Revision 2, Argonne Natl Lab., Chicago, IL (2016).

[58] ORGANISATION FOR ECONOMIC CO-OPERATION AND DEVELOPMENT/NUCLEAR ENERGY AGENCY, International Common-Cause Failure Data Exchange, ICDE General Coding Guidelines – Updated, NEA/CSNI/R(2011)12, OECD Publishing, Paris (2011).

[59] INTERNATIONAL ATOMIC ENERGY AGENCY, Methodologies for Seismic Safety Evaluation of Existing Nuclear Installations, IAEA Safety Reports Series No. 103, IAEA, Vienna (2020).

[60] NUCLEAR REGULATORY COMMISSION OFFICE ON NUCLEAR REGULATORY RESEARCH, Correlation of Seismic Performance in Similar SSCs (Structures, Systems, and Components), NUREG/CR-7237, US NRC, Washington, DC (2017).

[61] MCSWEENEY, L., "Seismic correlation modelling in multi-module PRAs" (Proc. Int. Topical Mtg on Probabilistic Safety Assessment and Analysis (PSA 2019), Charleston, SC, 2019), American Nuclear Society, La Grange Park, IL (2019).

[62] HENNEKE, D., LI, J., WARNER, M., "PRISM internal events PRA model development and results summary", in Proc. 13th Int. Probabilistic Safety Assessment and Management Conf. (PSAM13), Seoul, Republic of Korea, 2–7 Oct. 2016.

[63] CANADIAN NUCLEAR SAFETY COMMISSION, Probabilistic Safety Assessment (PSA) for Nuclear Power Plants, Regulatory Document REGDOC-2.4.2, CNSC, Ottawa, ON (2014).

[64] NUCLEAR REGULATORY COMMISSION, Procedure for Analysis of Common-Cause Failures in Probabilistic Safety Analysis, NUREG/CR-5801 (SAND91-7087), US NRC, Washington, DC (1993).

ANNEXES: COUNTRY REPORTS

The country reports presented in these annexes have been prepared from the original material as submitted by the contributors and have not been essentially modified by the staff of the IAEA. The country reports contain approaches and experiences of Member States (Canada, France, Hungary, the Republic of Korea, the United Kingdom and the United States of America) related to MUPSA. The IAEA is not responsible for the content of the Member State reports, and all questions must be directed to the individual authors or organizations.

Annex I

CANADA/COG — APPROACH TO MUPSA AT CANADIAN NPPs

I–1.   INTRODUCTION

This annex describes the typical approach used to prepare a PSA for a Canadian MU NPP. It includes a brief history of the evolution of PSA for Canadian MU NPPs and a description of the key features that affect the preparation and results of a PSA for a Canadian MU NPP. In addition, this annex discusses broader considerations on the assessment of whole site risk for an NPP site, including the topic of risk aggregation, along with some sample results.

I–2.   CANADIAN MU NPPS

All Canadian MU NPPs are located in the province of Ontario and are of the CANDU[1] design. There are three operating MU sites:

(1)   Bruce Power's site on Lake Huron, which includes two separate four-unit stations comprised of 4 × 831 MW(e) each;
(2)   Ontario Power Generation's Darlington site on Lake Ontario, which includes a four-unit station comprised of 4 × 881 MW(e);

---

[1] CANDU (Canada deuterium uranium) is a registered trademark of Atomic Energy of Canada Ltd, used under exclusive licence by Candu Energy Inc., a member of the SNC-Lavalin Group.

(3)   Ontario Power Generation's Pickering site on Lake Ontario, which includes
       a six-unit station comprised of 6 × 515 MW(e) (two additional units are in
       a safe storage state).

The design of the Canadian MU NPPs has evolved over time from the earliest units at Pickering to the most recently constructed units at Darlington. However, many of the key features that affect the preparation and results of a PSA are common to all three sites.

From the PSA perspective, a distinguishing feature of the Canadian MU NPPs is the degree to which SSCs are shared between the different units. For example, at the Darlington NPP, the four units share:

(a)   A common powerhouse;
(b)   A common containment envelope;
(c)   A common service water intake for post-accident service water supplies;
(d)   A common MCR;
(e)   A common supply system and a common recovery system for the emergency
       core cooling system;
(f)   Common post-accident electrical supplies;
(g)   Common post-accident service water pumps and piping.

Additionally, some systems have interunit ties that allow one unit to support another unit. For example, at the Darlington NPP, boiler feedwater, service water and instrument air can be supplied from a normally operating unit to a unit that is undergoing a transient.

As a result of the extensive sharing of safety related SSCs, Canadian MUPSAs have always addressed MU effects. The level of detail in addressing MU effects has increased with each successive PSA revision.

I–3.   EVOLUTION OF PSA

The first Canadian MUPSA was prepared in support of the design of the Darlington NPP. The Darlington probabilistic safety evaluation, completed in 1987, comprised a detailed Level 1 PSA for internal events for the reactor at full power, along with less detailed elements of Level 2 and Level 3 PSAs.

Between 1987 and 2006, PSAs with a similar level of detail to that of the Darlington probabilistic safety evaluation were prepared for the other Canadian MU NPPs.

In 2005, the CNSC issued its regulatory standard S-294 [I–1]. This regulatory standard introduced a number of requirements that drove changes to the Canadian MUPSAs. For example, S-294 included requirements to:

— Complete a Level 2 PSA;
— Update the PSA on a regular basis;[2]
— Seek CNSC acceptance of the PSA methodology;
— Include both internal and external events;
— Address both at-power and shutdown states.

The introduction of regulatory standard S-294 into the Canadian regulatory environment led to a major effort between 2008 and 2014 to update the existing PSAs.

The PSAs prepared between 2008 and 2014 were used to estimate both the severe CDF[3] (SCDF) and the LRF[4]. Both risk metrics were estimated on a per hazard, per unit basis and compared to safety goals on a per hazard, per unit basis.

The PSAs were used to determine the important contributors to risk and to identify opportunities to enhance NPP safety. In particular, if the estimated risk metric exceeded a safety goal, plans were developed to reduce the risk metric to a value below the safety goal through improved analysis, plant modification or a combination of both.

The PSAs prepared between 2008 and 2014 were also used to develop tools to support operational risk management and to prepare system specific reliability models to support the reliability and maintenance programmes.

Following the Fukushima Daiichi NPP accident, the CNSC replaced regulatory standard S-294 [I–1] with regulatory document REGDOC-2.4.2 [I–2]. This new regulatory document introduced a number of new requirements, such as:

— The requirement to include sources other than the reactor (e.g. the SFP);
— The requirement to include combinations of external hazards;
— The requirement to include operational states other than full power and shutdown.

The Canadian nuclear industry is working towards compliance with the requirements of REGDOC-2.4.2 [I–2].

---

[2] The current regulatory requirement is to update PSAs every five years.

[3] Severe core damage is defined as the failure of more than one fuel channel. The safety goal is 10^{-4} per reactor-year.

[4] A large release is defined as an atmospheric release of ^{137}Cs in excess of 10^{14} Bq. The safety goal is 10^{-5} per reactor-year.

I–4. PSA STRUCTURE

I–4.1. Hazard identification and screening

The first step in a MUPSA is to identify the internal and external hazards that, by themselves or in combination with the failure of mitigating systems, might lead to reactor damage. The list of potential hazards is then screened to eliminate those hazards that do not pose a challenge to the NPP.

The hazard identification and screening process generally follows that described in IAEA Safety Standards Series No. SSG-3 [I–3] and in ASME/ANS RA-Sb–2013 [I–4]. However, the PSAs also address combinations of external hazards, with particular attention paid to correlated hazards[5] and consequential hazards[6].

As a result of the hazard identification and screening process, detailed PSAs have typically been prepared for five hazards: internal events, internal fires, internal floods, seismic events and high winds.

I–4.1.1. Selecting a representative unit

The detailed PSAs are used to estimate SCDF and LRF on a per hazard, per unit basis.

To facilitate the estimation of a per unit risk metric, one of the units at the NPP is chosen as the reference unit and the risk metrics are estimated for that unit. As there are few design differences between the units in a Canadian MU NPP, the SCDF and the LRF for the reference unit are representative of the SCDF and the LRF for the other units.

At the Darlington NPP, for example, the reference unit is Unit 2.

I–4.2. Event tree and fault tree analysis

The Level 1 PSAs for Canadian MU NPPs are prepared using the small event tree, large fault tree approach. Event trees and fault trees are prepared following a typical industry approach [I–3, I–4].

---

[5] A correlated combination of hazards occurs when two hazards are parts of the same initiator, e.g. heavy rain and strong winds as part of a single weather system.

[6] A consequential combination of hazards occurs when one hazard directly causes another hazard, e.g. an earthquake causes a tsunami.

Each hazard that requires a PSA is broken down into a range of IEs. The list of IEs includes those that:

(a) Occur on the reference unit and affect only the reference unit (e.g. loss of reactor power control);
(b) Occur on an adjacent unit and affect the reference unit as well as the adjacent unit (e.g. steam from a large steam line failure on an adjacent unit causing an IE on the reference unit);
(c) Affect all units simultaneously (e.g. a LOOP, a MCR fire or a seismic event).

An event tree is prepared for each IE and a fault tree is prepared for each of the safety functions defined in the event tree. While the focus of the PSA is the reference unit, the event trees and the fault trees take into account MU dependencies, for example:

(1) A common IE can affect the reliability of the safety functions on all four units, as well as that of interunit safety functions. For example, failures associated with the common service water intake at the Darlington NPP can cause an IE and affect the reliability of the unitized, shared and interunit emergency service water supplies.
(2) The success criteria for the safety functions credited in the event trees take into account the number of units participating in the sequence. For example, more emergency service water pumps may be required to operate following an IE affecting four units than for an IE affecting a SU.
(3) The range of post-operator actions required to be performed in a sequence affecting four units might be greater than the range of actions required to be performed in a SU sequence. This might increase the probability of failure to perform the required actions, either as a result of increased complexity or increased time pressure.

The end state of the event tree is either a stable configuration of the reference unit or core damage in the reference unit.

I–4.3. Level 1 PSA solution

For each hazard, the sequences resulting in severe core damage in the reference unit are combined into a single-top fault tree. The solution to the single-top fault tree includes a hazard specific estimate of SCDF for the reference unit and a list of MCSs.

Through a detailed knowledge of the IEs and the accident sequences, the MCSs for the reference unit can be interrogated to identify single-, two- and four-unit sequences[7]. For example:

(a) Cut sets containing IEs that affect only the reference unit (e.g. a loss of reactor power control, are designated as SU sequences);
(b) Cut sets containing IEs that might affect more than one unit but contain unit specific mitigating system failures are designated as SU sequences;
(c) Cut sets containing IEs that might affect more than one unit and contain only failures of common mitigating system are designated as MU sequences;
(d) Cut sets initiated by a seismic event that contain only seismically induced failures of mitigating systems are designated as MU sequences[8].

Separating the MCSs into sequences that affect one, two or four units is a prerequisite for the Level 2 PSA and supports risk aggregation (see Section 5).

I–4.4. Level 2 PSA

In order to limit the Level 2 PSA to a manageable size, sequences that result in severe core damage in the Level 1 PSA are grouped according to their characteristics. The groups are called PDSs. The attributes of a PDS include:

— The number of units participating in the sequence;
— The type of IE;
— The shutdown status;
— The status of the reactor coolant system at the onset of core damage;
— The severity and timing of core damage;
— The mitigating system status.

One of the key characteristics of a PDS is the condition of the containment envelope; large releases do not typically occur for sequences in which the containment envelope remains intact. The containment envelope may:

(a) Be intact;
(b) Be bypassed by the IE, e.g. a steam generator tube rupture;

---

[7] Due to design considerations, three-unit sequences are very unlikely and are conservatively grouped with four-unit sequences.

[8] Seismic events are treated as being fully correlated. That is, the probability of seismically induced failures of unit specific equipment is assumed to be identical for all units.

(c) Be impaired as a result of independent failures, e.g. penetration failures, failure of an airlock's pneumatic door seals;

(d) Fail because of the failure of containment support systems[9];

(e) Fail consequentially because of phenomena[10] specific to the sequence.

Accident progression analysis using the Modular Accident Analysis Program for CANDU reactors (MAAP-CANDU)[11] is performed to support the grouping of sequences into PDSs. At the Darlington NPP, for example, the MAAP-CANDU analysis determined that:

(1) Unmitigated single-unit and two-unit sequences do not result in the consequential failure of the containment envelope and, therefore, do not result in a large release;

(2) Single-unit sequences in which the containment envelope is bypassed may result in a large release;

(3) Single-unit and two-unit sequences in which the containment envelope fails independently may result in a large release;

(4) Some single-unit and two-unit sequences in which containment support systems fail may result in a large release;

(5) Unmitigated three-unit and four-unit sequences result in the consequential failure of containment and a large release.

The LRF for the reference unit can be estimated by summing the frequencies of the sequences in the PDSs based upon items (2) to (5) above; for example,

(a) The frequency of sequences in the PDS based upon item (4) above can be estimated by combining the SCDF for single and two-unit sequences with the probability of failure of the containment support systems. The probability of failure of the containment support systems is estimated using fault trees.

(b) The frequency of sequences in the PDS based on item (5) above can be estimated by combining the SCDF for three- and four-unit sequences with the probability of the phenomena that cause the consequential failure of the containment envelope. The probability of the phenomena that cause

---

[9] Containment support systems include the reactor vault cooling system, the post-accident hydrogen ignition system and the filtered air discharge system.

[10] Phenomena include thermal hydraulic processes associated with reactor heat-up, melting and disassembly; ignition of combustible gases; energetic corium–coolant interactions; slow containment pressurization; and corium–concrete interactions.

[11] MAAP-CANDU is an integrated computer code for the best estimate analyses of severe accident scenarios in CANDU-type NPPs.

the consequential failure of the containment envelope is estimated by calculation or by expert judgement.

The solution of the Level 2 PSA includes a hazard specific estimate of the LRF for the reference unit and a list of MCSs. As explained in Section 4.4, Level 2 PSA can be interrogated to distinguish between sequences that affect one, two or four units, for use in supporting risk aggregation calculations (see Section 5).

IEs affecting more than one unit are the major contributor to the LRF at Canadian MU NPPs. This reflects the extensive sharing of safety related SSCs and, in particular, the shared containment envelope.

I–5. WHOLE SITE RISK CONSIDERATIONS

I–5.1. Background

In recent years, the term 'whole site risk' has become a subject of discussion at licensing hearings in Canada, given that the current PSAs are typically conducted on a per unit, per hazard basis, but most Canadian NPPs have multiple units on-site and may be subjected to different types of hazard. The discussion has centred around whether the safety assessments of MU NPPs account for the whole site risk, meaning the overall risk of the site due to multiple reactor units, other on-site sources of radioactivity (such as the SFPs and used fuel dry storage facilities), internal and external hazards, and other reactor operating modes (i.e. in addition to full power and shutdown states). As discussed in Section 5.2, whole site risk is evaluated through a combination of qualitative and quantitative factors covering normal operation, abnormal operational occurrences and accident conditions. As a supporting tool, whole site PSA contributes to an understanding of the broader topic of whole site risk. The key issues associated with whole site PSA include a lack of international consensus on the methodology; the appropriateness of risk aggregation across different hazards; and the acceptance criteria for a site based PSA assessment (for instance, there is no site LRF safety goal in Canada).

With respect to numerical risk aggregation, the simple addition of the per unit risks across all hazards may yield a biased result because of the large uncertainties and conservative assumptions associated with external hazards such as seismic and high winds (i.e. the sum of the means may not equal the mean of the sum). Furthermore, for a given hazard type, a MUPSA risk result (such as LRF) is generally not equal to the per unit risk value multiplied by the number of units on-site. Moreover, not all hazards are quantified in terms of PSA risk metrics (e.g. malicious acts), and hence, they do not lend themselves to risk aggregation by arithmetic summation of common risk metrics.

In response to this issue at the time, Ontario Power Generation committed to perform a whole site risk assessment for the Pickering NPP site by the end of 2017. In addition, the CANDU Owners Group (COG) published a concept level report on whole site PSA methodology [I–5], hosted an international workshop (January 2014) and participated in a CNSC-hosted international workshop (November 2014) on topics related to whole site PSA. Since then, a COG joint project was launched to further develop the concepts in the COG paper and to support the completion of the Pickering whole site risk assessment. The scope of the COG joint project includes work to:

— Facilitate communication on the evaluation of NPP safety and whole site risk (taking into account the role of PSA and the broader considerations that factor into risk informed decision making) [I–5];
— Quantify whole site PSA;
— Assess margins between LRF values and quantitative health objectives (QHOs);
— Address low power reactor operating modes;
— Address risk from non-reactor sources.

Highlights from some of the above areas are discussed in Sections I–5.2 to I–5.4 below.

I–5.2. Evaluation of NPP safety and risk

An objective of assessing whole site risk is to holistically demonstrate that a nuclear site with one or more reactors is adequately safe and does not pose an unreasonable risk to the surrounding population or the environment. In Canada (and in other jurisdictions), from the nuclear regulatory context, 'adequate safety' and 'reasonable risk' are value based judgements made by the authorized body. The Canadian utilities are responsible for operating NPPs in a manner that ensures adequate safety and reasonable risk, and the Commission Tribunal makes a determination of whether these objectives have been achieved. In its deliberations, the Commission Tribunal takes into account input from the utility, CNSC staff and other interested parties, including those who may be exposed to the risk, through the licence hearing intervention process. Judgements are based on a broad set of qualitative and quantitative information.

There is no requirement to reduce the input to the Commission's licensing decision to a single number. As such, whole site risk needs to be defined and evaluated in a way that supports the value based approach to safety determination and licensing decisions and takes into account both qualitative and quantitative information. That is, rather than attempting to characterize whole site risk by a single number or even a series of numbers, the approach needs to provide quantitative and qualitative information to support a judgement on whether the whole site risk is limited to a reasonable level. For this philosophical reason, together with the previously mentioned numerical issue of simple addition of risk results across all hazards, Canadian utilities do not support the summation of risks across all hazard types.

In practice, the CNSC staff evaluate the safety of NPPs across 14 safety and control areas (SCAs), comprising 73 subtopics. Each utility has formal programmes in place to ensure effective performance in each SCA. The SCA evaluations assess both qualitative and quantitative information and are reported annually for each NPP. These SCA evaluations consider actual experience, the fitness for service of equipment, the effectiveness of programmes and the safety analysis, including the results from both deterministic safety analysis and PSA. For example, recent evaluation results indicate that all Canadian NPPs performed at a 'satisfactory' or 'fully satisfactory' level for each SCA. In addition to the NPP ratings for each of the SCAs, there is an integrated plant rating for each NPP. A satisfactory or fully satisfactory integrated plant rating is an indication that the overall risk associated with each NPP site is limited to a reasonable level.

I-5.3. Whole site PSA

Further risk insights for MU NPP sites may be obtained through the calculation of PSA results on a station wide basis for each hazard type. This discussion focuses primarily on LRF, as it is a more direct indicator of risk to the public than SCDF. COG members have developed the following approach for risk aggregation of PSA results to express the LRF on a per station basis for a given hazard.

For a station comprising four reactor units, for example, the station LRF for a given hazard category is obtained as follows:

LRF per station = 4 × per unit LRF for IEs that affect a SU only.

+ 2 × per unit LRF for IEs that affect two units simultaneously (this term accounts for all possible combinations of two-unit events; here, the per unit LRF represents events involving the reference unit and one other unit).

+ 1 × per unit LRF for IEs that affect all units simultaneously (three-unit sequences are very few and are lumped in with four-unit cases).

The underlying basis for this formula is that IE frequencies are established on a per reactor basis independently of how many reactors are in a station. Therefore, if a random event affects only a single reactor unit, the probability that the event will occur in a MU station is the probability per reactor multiplied by the number of reactors. The other two terms account for events that can affect combinations of two, three or all four reactors at a time. The appropriate per unit LRF values for use in the above equation are obtained by interrogation of the cut sets from the PSA process described in Section 4 (i.e. to identify the single-, two-, three- and four-unit sequences involving a large release).

The following is a sample calculation for a four-unit station, where Unit 2 is the reference unit.

For the hazard category of high winds, the total per unit LRF is equal to 9.9×10^{-7}/year.

Cut sets from the high wind PSA were interrogated to identify single-unit, two-unit and four-unit sequences, as shown in Table I–1.

Thus, the per station LRF for high winds is equal to the following:
$4 \times 3.1 \times 10^{-7} + 2 \times 7.6 \times 10^{-9} + 6.7 \times 10^{-7} = 1.93 \times 10^{-6}$ per year

The same process is applied for each of the remaining hazard categories using the associated PSA elements. A tabular summary of the sample results is provided in Table I–2 below.

These sample results are also shown in the graph in Fig. I–1.

In this graph, the 'per unit LRF' lines represent the per unit based LRF values as calculated from each of the separate hazard PSAs, where MU effects were accounted for (in accordance with the discussion in Section 4). The set of SU-only data represents the portions of the per unit LRF values that are comprised of IEs that affect only a SU. The multi-unit data represent the per station LRF values as calculated via the above risk aggregation method.

The proximity of the per unit LRF line relative to the SU only LRF value illustrates the extent to which MU effects factor into the per unit LRF. For instance, as shown for high winds in this case, the per unit LRF line is well above the SU only value, which indicates that additional MU sequences contribute to

TABLE I–1. LRF RESULTS FOR DIFFERENT SEQUENCES

Sequence	LRF (per reactor-year)
Unit 2 only	3.1×10^{-7}
Unit 2 + one other unit	7.6×10^{-9}
Unit 2 + at least two other units	6.7×10^{-7}

TABLE I–2. LRF RESULTS FOR DIFFERENT HAZARD CATEGORIES

Hazard category	LRF	
	Per reactor ($\times 10^{-5}$/year)	Per station ($\times 10^{-5}$/year)
Internal events	0.10	0.20
Internal floods	0.02	0.02
High wind	0.10	0.19
Internal fires	0.08	0.28
Seismic	0.28	0.29
PSA safety goal	1.00	n.a [a]

[a] n.a: not applicable.

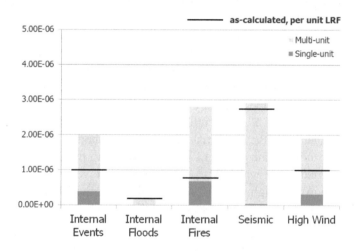

FIG. I–1. Example of site LRF calculations.

the per unit LRF (i.e. although the per unit LRF is 'per unit' based, it includes more than events involving the single (reference) unit only). In contrast, for the case of internal fires, the per unit LRF line is very close to the SU only value, which indicates that although some additional MU sequences are captured, they are not major contributors to the per unit LRF for fire (SU events dominate).

On a per station basis, the LRF includes a fuller account of MU effects (i.e. by consideration of LRF contributions directly from the non-reference units). The proximity of the 'per unit LRF' line relative to the site wide LRF value represents the extent to which the per unit LRF covers MU sequences across the station. For instance, as shown for seismic hazards in this case, the per unit LRF line is very close to the site wide value; that is, the per unit LRF largely encompasses the LRF calculated on a per station basis. In this case, it indicates that seismic risk is dominated by sequences where all units are simultaneously affected, as expected. Such a comparison of per unit LRF with per station LRF sheds light on the full extent of the inclusion of MU effects in the per unit LRF.

In essence, the above aggregation of per unit based LRF values across all units is a form of extrapolation. As such, it represents a pragmatic approach to estimating the site wide LRF for a given hazard type. It is noted that the approach requires careful decomposition of the per unit based PSA results and that some assumptions may be necessary in lieu of more detailed PSA modelling of all units. For Canadian MU NPPs, the per unit based PSA is in fact a multiunit PSA (MUPSA) as described in Section 4; however, this risk aggregation approach provides a more detailed MUPSA. Moreover, examination of the PSA results (i.e. SU only vs per unit vs 'per station' based) could provide useful information for risk insights and risk management.

I–5.4. Margins to QHOs

Fundamentally, as a universally accepted principle, an NPP has to pose only a small incremental risk to the life and health of the surrounding population. Depending on the jurisdiction, such a qualitative safety goal may be supported by one or more QHOs expressed in terms of a small percentage of the public health risk from other causes (e.g. the NRC's QHO related to latent cancer fatalities).

The LRF risk metric is considered to be a surrogate for QHOs. Meeting the per unit LRF safety goal for a reactor may be a sufficient condition to ensure that a QHO is also met. In Canada, a regulatory QHO is currently not required, given the comprehensive approach to assessing risk as described in Section 5.2. However, as part of the COG joint project, margins were assessed to possible QHOs for a range of LRF and magnitude assumptions. This involved a parametric study to delineate approximately at which postulated combinations of large release magnitudes and frequencies different QHOs would be just met, including for the aforementioned NRC QHO (for illustrative purposes). The preliminary results show that a total LRF in the order of 10^{-5} per year would provide margin to possible QHOs for a wide range of accident consequences. This suggests that a site wide LRF value may somewhat exceed 10^{-5} per year (which is the per unit/per hazard LRF safety goal in Canada) and yet still meet the overarching health objective. This order of magnitude for LRF may also serve as a reasonable basis for addressing other considerations associated with large releases of radioactivity from NPP accidents.

I–6.   SUMMARY

This annex has provided an overview of the PSA methodology for Canadian MU NPPs, as well as discussions and sample illustrations related to the topic of whole site risk for NPP sites. The key points are summarized as follows.

With respect to MUPSA in Canada:

— PSAs performed for Canadian MU NPPs have, by necessity (given the extensive sharing of SSCs), always been MUPSAs in that they explicitly account for MU interactions, even though PSA results are expressed on a per unit basis;
— Through careful risk aggregation, per unit based PSA results may be combined to fully quantify MUPSA (or whole site PSA) risk metrics for a given hazard type;
— The Canadian MUPSA approach could generally be used for other MU sites and reactor designs.

With respect to whole site risk:

— The evaluation of whole site risk involves the consideration of both qualitative and quantitative information that facilitates a value judgement of the reasonableness of risk and is informed by many factors within a broad perspective that includes various programmatic, deterministic and defence-in-depth considerations, as well as MUPSAs.
— PSA plays a complementary role to other factors and provides risk insights to improve plant safety.
— In Canada, whole site risk evaluation is supported by an integrated assessment using quantitative and qualitative information in 14 SCAs. Robust programmes are in place for each SCA to ensure effective risk management.

REFERENCES TO ANNEX I

[I–1] CANADIAN NUCLEAR SAFETY COMMISSION, Probabilistic Safety Assessment (PSA) for Nuclear Power Plants, Regulatory Standard S 294, CNSC, Ottawa, ON (2005).

[I–2] CANADIAN NUCLEAR SAFETY COMMISSION, Probabilistic Safety Assessment (PSA) for Nuclear Power Plants, Regulatory Document REGDOC 2.4.2, CNSC, Ottawa, ON (2014).

[I–3] INTERNATIONAL ATOMIC ENERGY AGENCY, Development and Application of Level 1 Probabilistic Safety Assessment for Nuclear Power Plants, IAEA Safety Standards Series No. SSG 3, IAEA, Vienna (2010).

[I–4] AMERICAN SOCIETY OF MECHANICAL ENGINEERS/AMERICAN NUCLEAR SOCIETY, Standard for Level 1/Large Early Release Frequency Probabilistic Risk Assessment for Nuclear Power Plant Applications, ASME/ANS RA Sb–2013, ASME, New York, NY/American Nuclear Society, La Grange Park, IL (2013).

[I–5] VECCHIARELLI, J., DINNIE, K., LUXAT, J., Development of a Whole Site PSA Methodology, COG 13 9034 R0, COG, Toronto, ON (2014).

FRANCE: IRSN — APPROACH TO MUPSA

II–1. INTRODUCTION

II–1.1. PSA in French safety demonstration

In the nuclear industry, France represents a unique situation, with a rather large fleet of NPPs (58 in operation) that are all built by the same manufacturer (AREVA NP) and operated by the same utility (Électricité de France — EDF).

Although the safety demonstration of the French NPPs is and remains deterministic, the probabilistic approach is taking an increasing place in the decision making process linked to safety. Despite a limited regulatory framework, the risk insights are more and more taken into consideration as a supplement to the traditional deterministic demonstration.

In order to clarify the acceptable approaches, the French nuclear safety authority (Autorité de sûreté nucléaire — ASN) requested the establishment of a Basic Safety Rule on the development and use of probabilistic safety assessments. The minimum scope of reference PSA, which is developed by the utility (EDF), is a Level 1 PSA, covering all internal IEs, as well as the loss of ultimate heat sink and LOOP for all POSs. The loss of ultimate heat sink and LOOP are IEs that are typically induced by external hazards. In order to be able to independently review the PSAs performed by EDF, the Institut de Radioprotection et de Sûreté Nucléaire (IRSN), ASN's technical support organization, develops its own PSA models.

II–1.2. External hazards

Worldwide operating experience shows that external hazards play an important role in the safety of nuclear facilities, because they have the potential to cause IEs and simultaneously to impair the safety systems necessary to limit the consequences of the IEs.

II–1.2.1.   External hazards in the safety demonstration in France

External hazards were initially considered as independent from internal incidents/accidents. Later, studies performed in the framework of the periodic safety reviews have revealed a correlation between some natural hazards and the loss of the heat sink or the loss of off-site electrical supplies. Consequently, the list of the natural hazards and combinations of hazards to be considered in the deterministic safety demonstration and the characterization of the 'design basis hazards' have been reviewed, considering mainly the operating experience, and many improvements of the protective measures have been implemented in all French NPPs.

Also, studies have been performed to analyse the types of accident that may affect all the site units at the same time and for a long duration. These studies have mainly brought about the following:

— Preventative warning requirements;
— Stronger requirements on the water inventory in the tanks necessary to refill the secondary water tanks of the auxiliary feedwater system in the case of a loss of the heat sink of long duration;
— Adaptation of accident procedures in order to deal with MU loss of the heat sink and of external electrical supplies;
— Improvement of on-site emergency planning to deal with MU accidents, in particular in the case of external hazards (with possible difficulties in accessing the site buildings);
— Diesel generator reliability improvements for long mission times;
— Identification of the need for additional probabilistic assessments for long LOOP (>24 hours).

Recently, additional design and organizational means resulting from the post-Fukushima 'stress test analyses' were studied:

— Additional equipment: ultimate diesel generator, additional pump to inject in the primary circuit, means to refill the secondary circuit, mobile ultimate means, additional ultimate heat sink (for reactor and fuel pool);
— Accidental procedures covering site accidents;
— Nuclear Rapid Action Force (FARN) — operational within 24 hours.

For the design of future plants, events that could affect several units on a site have been considered from the earliest stage of design. As an example, European pressurized reactor (EPR) safety demonstration requires that a SU is capable of coping with any relevant event without needing support from the neighbouring unit.

II–1.2.2.   Recent external hazards in France

In France, several external events with the potential to threaten nuclear safety have occurred. The most significant event was the partial flooding of the Blayais NPP in December 1999 during a severe storm, when high waves overtopped a protective dyke installed at the plant, partially submerging portions of the plant area. This event called into question the design basis used for the protection of NPPs against external flooding and the efficiency of the existing measures, especially the warning systems, site protection measures, protection of safety related equipment, procedures and emergency organization.

Some other representative external events have also affected French NPPs:

(a) December 2005 — Paluel site: ice formation on the grid transformers led to shutdown of all four reactors and isolation from the external power supply for a duration of less than 24 hours.

(b) December 2009 — Cruas Units 3 and 4: a significant amount of vegetation (around 50 m^3 compared with a monthly average of 5 m^3) blocked the water intake of the common pumping station of Cruas NPP Units 3 and 4, by clogging the filtration trash racks. Unit 4 was in a situation of total loss of the heat sink, and Unit 3 was in a situation of partial loss of the essential service water system.

(c) December 2009 — Fessenheim Unit 2: partial loss of the heat sink was caused by clogging of the filtering drum screens resulting from organic debris.

These incidents remind us that environmental conditions, changing over time, can impact the safety of nuclear reactors and highlight the need for better assessment of the risk related to external hazards, in particular by extending the scope of the PSA to include all relevant external events and combinations of external events. In this context, in France, both the operator (EDF) and IRSN work at improving methods to better take into account in the PSA the situations that might impact on more than one site installation (reactors or SFPs), in addition to other fields of work related to external events PSA (e.g. hazard screening analysis, long term accident sequence modelling). The Fukushima Daiichi accident confirmed the importance and imperativeness of such PSA developments.

II–2. TECHNICAL APPROACH TO MUPSA IN FRANCE

II–2.1. MU aspects in existing SUPSA in France

In French PSAs, the loss of the heat sink and the total loss of power supply affecting one unit have been studied in internal event PSAs since 1990. Even though these PSAs consider only one unit, some site aspects are already taken into account: twin units' mutual backup, shared site mitigations and shared resources. The PSA results may lead to plant modifications. For example, a modification (use of thermal inertia of refuelling water tank) was introduced to cope with the total loss of the heat sink and demonstrated its effectiveness during the Cruas 2009 event.

In IRSN's internal events PSAs, the total loss of the heat sink and total loss of power supply initiators are studied on the basis of the assumption that they affect all site units. Indeed, if the loss is caused by a natural hazard, IRSN assumes that all units may be affected. Based on past events, such an assumption, though rather conservative, does not seem over conservative: even if the cooling by the heat sink was totally lost at only one unit, two of the three other units would also be challenged (with partial loss of the heat sink).

In IRSN's internal events PSAs, this led to the following considerations:

(a) There is limited availability of water reserves for secondary cooling, due to the use of common reserves for several units and being designed to cope with a loss of the ultimate heat sink at only one unit;
(b) It is impossible for the common means on-site (such as the ultimate site diesel generator or other ultimate devices) to be used by more than one unit at the same time;
(c) There is an impact on the human factor, since only one safety engineer is available for twin units;
(d) It is impossible to use backup by twin unit specific systems (such as the charging line of the other unit, which can be used by the first unit as a substitute for safety injection in some situations).

EDF's external hazard PSAs (seismic and external flooding) also take into account the impact of the hazard on the site units by adapted modelling of the unavailability of site or shared mitigations and of the human factor.

II–2.2. MUPSA ongoing developments

II–2.2.1. IRSN MUPSA

IRSN considers that PSA could provide interesting insights into the sequences induced by a loss of the heat sink or the off-site electrical supplies caused by natural hazards and affecting the whole site. However, the development of IRSN's in-house PSA showed that modelling improvements are necessary.

The IRSN MUPSA's ongoing developments are integrated into a larger activity intended to develop methods and tools for the extension of the PSA scope (mainly to external hazards), to include also the modelling of long term accident sequences, post-Fukushima mitigations and FARN. IRSN is also involved in the OECD/NEA Working Group on Risk Assessment Task Group on MUPSA.

The approach that is intended by IRSN is to extend the unit PSA by taking into account the impacts that a situation affecting other site installations (reactors or SFP) might have: shared SSC, shared reserves, human factor (specific procedures, higher stress, higher complexity of actions, errors of commission), management of multiple unit IEs, complications due to core damage on other units and timing of event propagation. The role of post-Fukushima features may also be highlighted.

In IRSN's opinion, the SUPSA, complemented with site aspects, can provide useful insights to help evaluate the site's capability to cope with events that may lead to multiple unit IEs or multiple core damage. In general, IRSN is less interested in the fulfilment of criteria on the absolute values of risk (no probabilistic safety objectives are defined in France) than in identifying safety improvements, such as the following:

(a) Identification of important 'shared' site vulnerabilities in the case of external hazards;
(b) Assessment of the sufficiency of site equipment and resources and of accident guidance to cope with site events;
(c) Identification of mitigation strategies (for example, priority to the use of shared equipment, taking into account the kinetic and the severity of the accident on different units);
(d) Assessment of the advantages and inconveniences of sharing systems and/ or resources between site units (mutual help for some events, reduction of mitigation capabilities for some other events).

II–2.2.2. EDF MUPSA

IRSN has recommended that EDF consider the development of MUPSA since early 2000. In the framework of the fourth periodic safety review of its 900 MW(e) plants, EDF committed to developing probabilistic methods for the study of loss of heat sink and/or off-site power that might affect more than one site installation, including the modelling of post-Fukushima mitigations. In order to meet this commitment, EDF developed a methodology and a test study, which analyses the impact of site events on:

— IEs;
— Mitigations, including interunit CCFs;
— The human factor.

The test study, which is based on a SU Level 1 'internal events' PSA, considers the reactor and the SFP for a site with two units (each unit: one reactor and one SFP). The objective is to evaluate the risk increase of considering site events compared to SUPSA.

Two types of IE are distinguished:

(a) IEs that affect only one unit. The site event can be produced only by the independent occurrence of such an event on two units. This situation is considered as having a negligible probability and is therefore not further studied.
(b) IEs that have the potential to lead to core damage on more than one unit.

Based on unit PSA, these IEs can be:

— Site long LOOP supply;
— Loss of heat sink affecting site units.

As a first evaluation, the frequency of site IEs is extrapolated from the unit IE frequency considering a 1/3 conditional probability of having a site event.

The mitigation resources (material or human) are categorized as follows:

(a) Based on use:
    (i)   Individual (can be used by only one unit);
    (ii) Shared (can be used simultaneously by two units).
(b) Based on belonging:
    (i)   Unit;
    (ii) Site.

The accident sequences that involve the use of individual resources belonging to the site, as well as those involving the use of shared resources (if autonomy is not sufficient for two units), might be affected by the consideration of site accidents.

In order to evaluate the interunit CCF between important individual resources, a factor of 1/3 on unit CCF was considered. The interunit CCF is important for mitigation of those failures that occur before the use of individual resources belonging to the site (as these resources cannot be used by more than one unit).

Regarding the probabilistic assessment of the human factor, a specific approach was proposed, based on the use and belonging of resources (MCR operators, safety engineers, field operators and crisis teams) and on the unit PSA HRA quantifications.

First, for each IE the following approach was applied:

— Identification of mitigation resources on the MCSs of SUPSA;
— Reassessment of mitigation resource failure probabilities taking into account the site event;
— Quantification of MCSs with the reassessed failure probabilities.

Second, the post-Fukushima features were also considered when applicable.

The results of the study indicate that the main contributions to risk increase are:

(1) Failure of unit resources by interunit CCF when the use of individual site resources is necessary;
(2) Human factor (especially for shutdown states).

The study was performed by considering the two units at full power; the most likely situation, according to EDF. Moreover, considering two units in the same state appears more penalizing, insofar as the same mitigation means are mobilized, following a similar chronology. For SFP scenarios, all POSs were considered, including refuelling shutdown and unloaded reactor states, which, although less likely, concentrate the risk.

The study showed globally that there is no cliff edge effect on unit 'internal event' Level 1 PSA results if the site events are considered. The use of post-Fukushima mitigations might have an important impact on the risk reduction.

The observed increase in risk (about 10–20%) nevertheless underlines the importance of considering the MU aspects for situations that may affect several facilities at the same site, in order to improve the PSA insights.

The EDF methodology and test study were analysed by IRSN in the framework of its fourth safety review of 900 MW(e) plants. IRSN concluded that the proposed methodology is acceptable with some minor additions. It was agreed between IRSN and EDF that, based on this methodology, the site aspects will be appropriately taken into account in future PSAs (fourth safety review of 1300 MW(e) plants).

Annex III

HUNGARY/NUBIKI — DEVELOPING A SITE RISK MODEL FOR THE PAKS NPP

III–1. INTRODUCTION

In accordance with the nuclear safety regulations in Hungary, the total risk originating from the operation of a NPP needs to be adequately quantified. Also, for MU sites, interactions between plant units need to be considered in the justification of safety. Site level risk assessment is seen as necessary to meet these requirements, as opposed to assessing risk separately for each unit at a MU site.

As a first step in the preparation for assessing site risk, a study was conducted to examine the feasibility of developing a site risk model and quantifying the site level risk for the four VVER-440/213 type reactor units of the Paks NPP in Hungary, based primarily on the use of the existing unit specific PSA models. A small scale analysis was subsequently performed for the LOOP IE to further research the risk modelling and quantification options outlined in the feasibility study. A full scope Level 1 PSA for the Paks site is now in preparation, making use of the achievements of the preparatory analyses performed so far.

III–2. OBJECTIVES

The ultimate goal of the analysis is to determine the site level risk attributable to the operation of four VVER-440/213 reactor units at the Paks NPP. Construction of a MUPSA model aims to achieve this goal. The most important value expected from the analysis is not only the risk quantification in itself, but an improved understanding of plant vulnerabilities to events that can challenge multiple plant units simultaneously. This understanding can help to better evaluate the effectiveness of plant design solutions and safety upgrades that have been supported by considerations of risk primarily on the basis of SU assessments.

III–3. SCOPE

The scope of the site risk assessment for NPP Paks can be described by addressing the following scope attributes:

— Level of analysis;

— Release sources;
— IEs;
— POSs.

Nuclear safety regulations require Level 1 as well as Level 2 PSA for NPPs in Hungary. On these grounds, a site Level 2 PSA is seen as necessary, because in comparison to SU analyses, it can result in a better characterization of release magnitudes and release frequencies, which enables a refined description of the health and environmental consequences of severe accidents at a MU site. A graded approach is followed in the site level risk assessment for NPP Paks. Accordingly, the current efforts are devoted to developing a MUPSA model that can be used to determine the frequencies of single and multisource severe accidents (core damage and fuel damage frequencies). Thus, this phase of the analysis is concerned with Level 1 PSA, keeping in mind the requirement to subsequently develop the assessment to a Level 2 PSA.

The release sources to be addressed in the analysis cover the four reactor cores and the fuel elements stored in the four spent fuel storage pools adjacent to the reactors in the reactor hall. Accidental releases from other sources of radioactivity have been assessed negligible in comparison to the releases from reactor core or spent fuel damage accidents and, accordingly, other sources are not included in the assessment.

The study is expected to cover all those IEs that have been subject to analysis in the plant specific, SU Level 1 PSA for the Paks plant. These IEs include internal events, internal hazards and external hazards screened in for quantitative risk assessment (see Section IV–3 for more details).

Plant operation at full power, as well as low power and shutdown states of partial and total refuelling outages, are in the scope of the analysis.

In summary, an 'all sources', 'all hazards' and 'all modes' Level 1 PSA is aimed at in the current phase of site risk assessment and the associated MUPSA for NPP Paks.

III–4. APPROACH

The feasibility study included a review of risk measures applicable to quantifying site level risk, focusing mostly on Level 1 PSA measures with some discussion on Level 2 PSA aspects. Combined POSs of the four reactors and the adjacent SFPs were characterized using the distinct POS defined for the unit level PSA models. The modelling needs of different types of IE in a site level analysis were identified. Most importantly, approaches seen as viable for assessing site level risk were discussed and evaluated. A preliminary MUPSA model of the

LOOP IE was then constructed for Units 1 and 2 of the plant by experimenting with the analysis approaches outlined in the feasibility study. Besides the use of common PSA methods, the analysis also included some developmental work for risk quantification software.

The feasibility of a site level risk analysis for the Paks NPP was assessed by reviewing publicly available information on international experience in this analysis area and giving consideration to the specifics of the existing unit level PSA studies and models for the plant. The review of international experience covered the activities of the IAEA [III–1], the conclusions of an international workshop organized by the CNSC on this subject [III–2] and the interim findings of the European ASAMPSA_E (Advanced Safety Assessment Methodologies: Extended PSA) research project (http://asampsa.eu). Since it was found that site level risk assessment and the associated MUPSA were still in the early phase of development worldwide, good practices could not be identified. Thus, the feasibility assessment had to be based mostly on the judgement of the analysts. Particular emphasis was placed in the study on the following, largely interrelated aspects:

(a) Metrics applicable to describing risk at a MU site;
(b) Definition of site level POS with consideration of multiple sources of release;
(c) Selection of IEs important to modelling MU effects;
(d) Modelling of concurrent (combined) MU or/and multisource accident sequences;
(e) Description of human reliability in the case of a MU or multisource accident;
(f) Modelling and quantification techniques.

The pilot exercise on the LOOP IE was performed to experiment with some of the analysis options outlined in the feasibility study. The findings from these two preparatory steps have been used to specify the tasks to be performed in site level risk assessment.

III–4.1. Risk metrics

The Level 1 PSA for NPP Paks includes the quantification of CDF in the reactor PSA and fuel damage frequency in the SFP PSA separately for each of the four units. In principle, the frequency of single and multiple core damage and fuel damage sequences has to be known and aggregated correctly to quantify risk at site level. It is noted that fuel damage can be regarded as a generic term and core damage sequences, quantified commonly in a Level 1 PSA, represent a subset of the entire space of fuel damage situations. However, for the sake of simplicity, let us just consider core damage to indicate the measures that can be used for quantifying plant risk. This formalism can then be easily extended to severe accidents of potential release sources other than the reactors at a site, including SFPs in particular.

The SCDF for a four-unit site like Paks can be expressed as given in Eq. (III–1):

$$SCDF = \sum_{i} CDF_i + \sum_{\substack{i,j \\ i<j}} CDF_{ij} + \sum_{\substack{i,j,k \\ i<j<k}} CDF_{ijk} + CDF_{1234} \qquad \text{(III–1)}$$

where

CDF_i is the cumulative annual frequency of accident sequences leading to core damage at one unit out of four,

CDF_{ij} is the cumulative annual frequency of accident sequences leading to core damage at exactly two units (i and j) out of four ($i < j$),

CDF_{ijk} is the cumulative annual frequency of accident sequences leading to core damage at exactly three units (i, j and k) out of four ($i < j < k$),

CDF_{1234} is the cumulative annual frequency of accident sequences leading to core damage at all four units.

The value of CDF_i cannot be precisely quantified by using the existing unit specific analyses. This is partly because some of the core damage sequences included in the unit specific PSA models may overlap, i.e. some portion of the unit specific CDF expressions may be attributable to multiple core damage sequences. In addition, there can be transients affecting more than one unit at a time, although core damage occurs at one unit only. Such scenarios have not been fully analysed in the unit specific analyses. The quantification of terms CDF_{ij}, CDF_{ijk} and CDF_{1234} assumes the development of a multireactor risk model.

For the purpose of Level 2 PSA, the frequency of large releases from single as well as multiple sources has to be determined in order to quantify risk at site level. If there is a single end state of the Level 2 analyses, e.g. large release or large early release, then the frequency of site level release (site LRF or LERF) can be obtained by using a formalism similar to that applied to core damage in Eq. (III–1). Accordingly, 15 combinations of large releases from the four reactors have to be considered for the Paks plant. By taking large releases from the SFPs into account, the number of release combinations increases up to $2^8 - 1 = 255$. It is emphasized that there can be serious differences between the consequences

150

of the different release combinations. From the perspective of Level 2 PSA, the magnitude of release associated with combined releases is also of concern, not merely an estimate on the overall LRF. This aspect is of particular importance if one intends to take account of environmental consequences (e.g. implications for Level 3 PSA). In the Level 2 PSA for NPP Paks there are 15 source term groups for reactor accidents and two source term groups for SFP accidents. For the four reactors and for the four SFPs, the number of source term group combinations is in the order of 10^6 if source term groups used in the unit specific Level 2 PSA are combined mechanistically for multiple sources of release. This is not manageable in practice; therefore, the feasibility study suggested that a limited number of site level release groups needs to be defined, as opposed to literally combining source term groups applied in the unit specific PSA for a SU. Since the current analysis phase focuses on Level 1 PSA, no considerations have been given yet to the definition of such release groups.

III–4.2. POS

There are 25 POSs in the reactor PSA for a SU of the Paks plant. These states cover full power and 24 low power and shutdown states representing refuelling outages. The operational states of the SFP are decomposed into six categories in the PSA, based on the level of decay heat, the number and storage configuration of fuel assemblies and the water inventory (normal operational level and refuelling level) of the pool.

In the analysis of an IE that impacts on multiple units or release sources, the operational state of the four reactors and the four SFPs at the time of the event has to be taken into account. For example, an IE can find the plant in such a state that the reactors of Units 1 to 3 operate at full power (POS No. 0 in the reactor PSA) with the corresponding SFPs characterized by a normal operational volume of water inventory and a low level of decay heat (POS No. 5 or No. 6 in the SFP PSA), while the fourth reactor is subject to refuelling (POS No. 10 in the reactor PSA), its SFP is filled up to refuelling level and the decay heat is medium-high in the pool (POS No. 2 in the SFP PSA). The combined states of the different release sources are called the overall POS.

In order to define the overall POS, the operational cycles of the four reactors and the four SFPs have been evaluated for a ten year period between August 2017 and July 2027, based on outage planning. The plant has recently introduced a 15 month operational cycle instead of the earlier 12 month cycle. Three types of refuelling outage — short, medium and long — are used in the new cycle. The periodicity of the cycle for a plant unit is ten years. The evaluation has led to the definition of 115 distinct overall POSs. Each overall state is characterized by a unique and physically viable combination of operational states for the four reactors and four SFPs. The durations of these overall POSs have been normalized so that they sum up to a year (8760 hours).

Figure III–1 exemplifies the nature of the overall POS for a 15 month (10 950 hour) operational cycle following the long outage of Unit 1 in 2017. The figure also shows the different states of the reactors and the SFPs.

In principle, a submodel within the MU risk model has to be developed for each overall plant state to appropriately represent the distinguishing characteristics of a state. In practice, it may be possible to reduce the number of overall plant states based on further, comparative analyses of the states to find bounding plant states for some groups of states (e.g. merging states with similar operational features, bounding low frequency states with less favourable states).

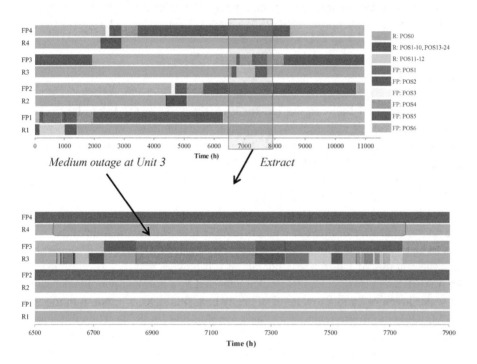

FIG. III–1. PSA based POS for a 15 month operating cycle for four reactors and four SFPs at NPP Paks.

III–4.3. IEs

The internal events PSA for the reactors of the Paks plant includes 70 IEs grouped into 14 categories, as follows:

(A) Reactor vessel damage, 4 IEs;
(B) Large LOCA, 8 IEs;
(C) Medium LOCA, 11 IEs;
(D) Small LOCA, 3 IEs;
(E) Interfacing system LOCA (including primary to secondary leaks), 6 IEs;
(F) Decrease in primary coolant flow, 2 IEs;
(G) Loss or reduction of feedwater flow, 9 IEs;
(H) Decrease in steam flow, 2 IEs;
(I) Loss of steam, 4 IEs;
(J) Transients causing turbine trip, 4 IEs;
(K) Electric power supply and instrumentation and control faults, 5 IEs;
(L) Support system failures and common cause initiators, 4 IEs;
(M) Unplanned reactor trip, 1 IE;
(N) Reactivity induced transients, 7 IEs.

LOCAs (IE groups A to E) impact only on a SU directly. Therefore, there is no need for a MU model as far as the direct consequences of these events are concerned. However, a LOCA at a unit can indirectly lead to transients at other units if there are severe consequences of a LOCA induced event sequence outside the boundaries of the affected unit. Forced shutdown of the neighbouring units due to the radiological impact of a LOCA initiated severe accident is an example of indirect effects that require consideration during the development of a site risk model. Such domino effects need, in general, to be taken into consideration for any type of IE included in the unit specific PSA. The limiting conditions of operation, the actual as well as the foreseeable consequences of an accident at a unit, determine the required response to the accident at other units. The decision on the response needs to be made by a responsible person or team. It can be the unit shift supervisor, the shift supervisor (common for the four units) or the emergency response team, depending on the status of accident progression.

Most of the transient IEs included in groups G to N also impact only on a SU directly, which suggests that a consequential IE at other units is not to be expected. However, it is to be examined whether high energy feedwater and steam line breaks within IE groups G and I can lead to MU transients as a consequence of intersystem interactions in the turbine hall, or not. Thanks to the capabilities of the applied modelling tools, these transients can be identified and evaluated on the basis of the existing internal flooding PSA for the Paks plant, as discussed further below.

Loss of power supply to the three 6 kV safety buses of a unit due to on-site failures is described by IE K1_B in the Paks PSA. A fault tree has been developed for this IE that includes various types of electrical component failure at a unit and in the switchyard that lead to loss of power without LOOP. The feasibility study found that the fault tree available for all four units was appropriate for identifying SU as well as MU on-site power failure events. The MU power supply failures are then to be the subject of MUPSA modelling.

IE K1_K is LOOP in the unit specific PSA model for Paks. This event affects the whole site. Although the PSA model for the K1_K IE takes the possibility of ensuring power supply to a unit in island mode of operation into account, this option is considered individually for each unit in the unit specific analyses. For a more realistic characterization of handling an off-site power event, a MUPSA model is required.

PSA modelling of internal hazards covers internal fires and internal flooding for Paks. The internal hazards PSA was largely supported by a dedicated database and analysis system. This analysis tool can be used to determine the consequences of a fire or internal flood event in terms of induced failures of SSC that cause a plant transient and degradations in mitigating systems. Relevant data for all four units are included in this database and analysis system, which makes it appropriate for determining MU fire or flood induced transients, and their consequences too.

The PSA model for external events covers seismic events, high winds and extreme snow, as well as ice formation (glaze ice and frost) and external events endangering water intake from the river Danube. Numerous other external events were considered in the analysis; most of them were screened out from detailed modelling, but follow-on analysis is still ongoing for a range of external hazards. Unlike internal events and internal hazards, the PSA for external events is currently available for a reference unit of the plant and not for each unit. External events included in the Paks PSA typically impact on the whole site. The feasibility study stressed the need for developing a site level MUPSA model for these events to fully quantify risk. In effect, external events are the focus of attention in a site level risk analysis.

The above statements on the role of the different types of IE in site level risk modelling are equally valid for the reactor and for the SFP, respectively. However, the transients that can lead to fuel damage in the SFP are limited to loss of cooling and LOCAs. These transients can result from either external or internal failure causes. (For the sake of completeness, fuel damage caused by a direct impact, such as large structural damage, needs to be mentioned too.)

In summary, the feasibility study found that the following categories of IEs need to be subject to modelling MU (and multisource) effects in PSA:

(a) Loss of power due to on-site causes;
(b) Loss of off-site power;
(c) Internal hazards included in the SUPSA — fire and internal flooding;
(d) All external hazards included in the SUPSA — seismic, high winds, extreme snow, ice formation, external events endangering cooling water intake and others;
(e) Any SUIE that indirectly causes a transient (e.g. forced shutdown) at other units (domino effect).

The LOOP event was the subject of the small scale trial analysis for two reactor units of NPP Paks.

III–4.4. Major analysis tasks in modelling MU accident sequences

In general, for all those categories of IEs that are in the scope of modelling MU effects, multiple transients affecting some of the four reactors and the four SFPs need to be identified, and the responses of the reactor and SFP systems as well as the operating personnel (including successes and failures) need to be modelled and quantified in an integrated manner in a MUPSA. The feasibility study included a review of the most demanding and crucial tasks in developing such a model. Below is a concise description of some important conclusions of the review, using also the insights from the small scale trial analysis.

III–4.4.1. Approach to modelling

Two basic options were studied and evaluated for the purposes of modelling and quantifying site level risk:

— Option 1: Event tree linking;
— Option 2: MCS conjunction.

Option 1 is the interconnection of the unit level accident sequences for each IE that can lead to a transient in more than one unit or release source. Interconnection can be made by building a single large event tree that includes all the combined event trees of the four units or by connecting a continuing event tree built for a unit to each event sequence (to success as well as to failure sequences) of another unit.

Option 2 is conjunction and subsequent Boolean reduction and quantification of unit level MCSs generated for a given end state (core damage or fuel damage) for an IE that induces transients at multiple units.

III–4.4.2. Concise review of options

(1)   Event tree linking

A small scale example of a combined single event tree for core damage is shown in Fig. III–2 for an IE that affects Units 1 and 2 simultaneously. The event tree end states are:

— S: success (no core damage at any plant units);
— CD1S2: core damage at Unit 1 only;
— S1CD2: core damage at Unit 2 only;
— CD12: core damage at Unit 1 and Unit 2.

The complexity of the combined event tree increases progressively as more units and more release sources are taken into account. If the individual event trees to be combined include n_i headers and m_i accident sequences for reactor i, then this setting will lead to a combined event tree with Σn_i headers and $\prod m_i$ sequences ($i = 1, 2, 3, 4$). It is not practical, nor is it even possible, to construct such large event trees manually. The use of continuing event trees does not simplify the modelling solution either, because the number of continuing event trees needs to be multiplied according the number of end state combinations

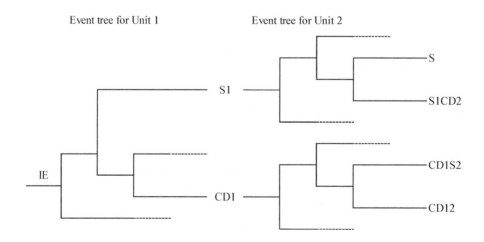

FIG. III–2. Construction of a single, large, combined event tree to model a MU event.

(single and multiple successes or failures for the four units) to enable a correct description of the consequences. It is therefore suggested to develop and use dedicated software for model construction in both cases.

Another method of event tree linking is the conversion of all the core damage sequences of the event tree for the relevant IEs at a unit into a fault tree. This can be done by building a fault tree representation of each core damage sequence and connecting these fault trees under an OR gate. Figure III–3 illustrates this modelling method for the same problem depicted previously in Fig. III–2. The complement of successful response at a unit is modelled by a fault tree conversion of core damage sequences in the fault trees linked to the headers of the event tree. This solution does not result in a large event tree or numerous event trees, and additionally, it can be done manually by using traditional PSA software. However, the complexity of the fault trees increases greatly.

(2)   MCS conjunction

This approach assumes the generation of MCSs for each unit and each release source separately and the subsequent combination of those cut sets, rather than the development of an integrated model. Quantification of any end state combination is straightforward, and unlike event tree linking, the method can also be relatively easily extended to Level 2 PSA. On the other hand, this approach has some weaknesses as well. Success branches are not represented in the MCSs; therefore, end state combinations that include success sequences cannot be quantified properly. For example, this method can be used to determine the frequency of CD12, but not CD1S2 or S1CD2 of Figs III–2 or III–3. The precision of the solution is substantially affected by the number of MCSs retained for the analysis and the goodness of the approximation cannot be assessed correctly.

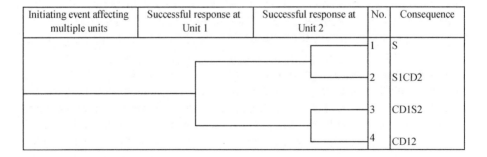

Initiating event affecting multiple units	Successful response at Unit 1	Successful response at Unit 2	No.	Consequence
			1	S
			2	S1CD2
			3	CD1S2
			4	CD12

FIG. III–3. Simplified combined event tree using fault tree conversion of accident sequences.

III–4.4.3. Choice and rationale

The general PSA modelling tasks have to be performed, irrespective of the modelling options used. Thus, the combination of the unit level models (either at the level of accident sequences or MCSs) can only yield meaningful results after the models to be combined are prepared for capturing MU effects and phenomena.

A small scale pilot study was prepared to experiment with the different modelling options on a real example. Loss of off-site power (IE K1_K) was selected for the purpose of the pilot, mainly because it is a MUIE itself, and a likely consequence of external events that are important to site risk is this IE too. To limit the size of the problem, the study was restricted to the reactors of Units 1 and 2 operating at full power at the time of the IE.

A PSA model for the two-unit analysis was developed by modifying the event tree models of the K1_K IE in the SUPSAs of the two units to (a) accommodate MU effects specific to off-site power and (b) enable a logically correct analysis by appropriate treatment and designation of shared, as well as strictly unit specific, systems and other model elements. Most importantly, the electric power supply configurations with the associated power transfer possibilities between the units following a K1_K event were studied and evaluated in cooperation between PSA experts and the operating personnel. Three power transfer modes were modelled, one of them being power transfer from a unit operating in house load to its twin unit via the so-called backup power buses, as exemplified in Fig. III–4. It is noted that a full scope HRA was beyond the scope of the study; therefore, only rough initial estimates were used to describe the probability of successfully establishing an appropriate power supply configuration in a given power supply fault scenario.

The two-unit PSA model was prepared to enable risk quantification by the use of both options envisaged in the feasibility study.

The event tree linking approach was applied to combine the prepared K1_K event tree of Unit 1 (denoted by 1_00_K1_K in the model, with the first character '1' identifying Unit 1 and '00' indicating full power) with the same kind of event tree for Unit 2 (denoted by 2_00_K1_K in the model), in accordance with the structure shown in Fig. III–2. In addition to directly interconnecting all the sequences of the 1_00_K1_K and the 2_00_K1_K event trees to create a large combined event tree, all the core damage sequences of both the 1_00_K1_K and the 2_00_K1_K event trees were converted into fault trees. The resulting large fault tree for 1_00_K1_K was linked to the first header, while the fault tree for 2_00_K1_K was linked to the second header of the combined event tree, in agreement with the model structure depicted in Fig. III–3. To be logically correct, the success branches of the core damage sequences were also modelled in the converted fault trees, not only the failed operations. The third method, i.e. the

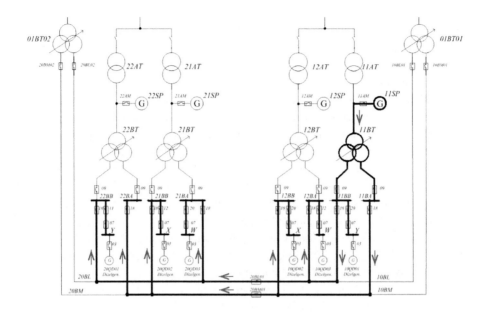

FIG. III–4. Power transfer between two twin units via backup buses.

use of continuing event trees, was not tested in the pilot because it could have actually been the same model as the combined large event tree, but split into a set of continuing event trees with identical end states.

MCS conjunction was applied by generating the MCSs for the core damage sequences of the 1_00_K1_K and 2_00_K1_K event trees and subsequently 'multiplying' these cut sets. The MCSs for the individual event trees and accident sequences were produced by the use of the RiskSpectrum PSA code. Dedicated software was developed and applied to perform Boolean reduction and quantification of the combined cut sets.

The conclusions of the feasibility study and the findings of the small scale pilot assessment were used to determine the advantages and disadvantages of the different modelling options. By comparing these advantages and disadvantages, use of the event tree linking approach with fault tree conversion of accident sequences was proposed for the purposes of the full scale MU Level 1 PSA of NPP Paks.

III–4.4.4. Shared resources

The shared resources that may have an important role in MU accident scenarios are basically plant design (technology) and human related. Technology related shared resources are discussed in this section; the human related ones are addressed in Section III–4.4.6.

Because of the design features of the Paks plant, there are resources that are common to multiple plant units. An example is the demineralized water system that is shared by two units. Open loop cooling by steam dump to the atmosphere is required for successful secondary side heat removal in some accident sequences, and demineralized water needs to be injected into the steam generators in this situation. If the inventory of the demineralized water tanks decreases below the limit prescribed in the conditions of operation, then the twin unit has to be shut down in accordance with the limiting conditions of operation. Thus, a reactor trip transient occurs at the twin unit. This combined scenario is not modelled in the SUPSA, but it has to be considered in the MU model.

The feasibility study identified 16 categories of shared technical resources for the four Paks units, including shared systems, shared structures and shared plant areas outside the building enclosures.

Both the success and the core damage sequences of the SUPSA models need to be reviewed to determine whether the use of shared resources can cause a transient or require interventions that lead to a transient at other units. If such combined events are to be accounted for, they need to be modelled together in the MUPSA by giving appropriate consideration to the reduced availability of shared resources.

A much limited subset of shared resources was considered in the small scale analysis. In this respect, most of the analysis tasks were left for the full scale assessment.

III–4.4.5. Correlated failures

External events typically lead to multiple failure events at the plant. These failure events are correlated to some extent. Induced correlated failures may occur not only within the boundaries of a SU, but also in multiple units. For example, the study pointed out that, in accordance with the assumptions and results of the fragility analysis in the Paks seismic PSA, not only do the seismic accident sequences of the reactor and the SFP overlap for a unit, but a number of fragility groups used in the SU seismic PSA need to be extended to actually include the relevant SSC of all four units. This approach assumes full correlation of numerous interunit seismic failures, which suggests that the cumulative frequency of MU seismic induced core damage sequences is close to the SUCDF. An example of plant wide seismic failures is damage to the large turbine building complex. This building complex was handled as a single component in the fragility analysis, and the probability of total building collapse was assessed. Accordingly, a seismic failure event that is common to the four units needs to be included in the MU seismic PSA, and the consequential component and system failures have to be identified at each unit to describe the impact of turbine hall failure. If one intends to relax the rigorous, and presumably overly conservative, assumption of full correlation, then substantial additional analyses have to be performed, and refinements have to be made to the fragility analyses available for the plant today. Similar considerations hold for other external events too, although the level of correlation among multiple failures depends strongly on the type of external event and the associated loads, as well as on the types and modes of induced component and system failures.

Dependent failures within the broader category of correlated failure events will, in general, be treated similarly in a SU and in a MUPSA. The overriding principle is to explicitly address dependence between failure events to the greatest extent possible for the various categories of dependence, including physical dependence, functional dependence and dependence between HFEs. For functional and human related dependence, it is necessary to consider and model resources that are shared, or need to be shared, when combating MU or multisource accident sequences at a time. These aspects are briefly discussed in separate sections in this annex.

CCFs as residual failure events not modelled explicitly in PSA can theoretically be extended to CCFs of components belonging to different units. However, the parametric models used generally to describe and quantify CCFs do not seem readily applicable to MU CCF events. To this end, it is noted that even intersystem CCFs or CCFs of a large number of components are rarely addressed in contemporary PSA. The feasibility study did not endeavour to propose a method to overcome this shortcoming.

III–4.4.6. HRA

As modelled in the Level 1 PSA for Paks, the responses of the plant personnel to plant accidents are governed by the symptom-orientated emergency operating procedures, and the responses are decided and taken by the operators separately for each unit. This is mostly true for MU accidents too, but it is the responsibility of the shift supervisor to decide on the use of shared resources. This decision, and the associated human related dependence, will be included and quantified in the MUPSA.

The effectiveness of accident management actions is the focus of Level 2 PSA. In the case of a severe accident, there is a shift in decision from the MCR personnel to the emergency response team supported by the TSC. The technical support staff consult the SAM guidelines to aid decision making. (The emergency operating procedures include exit points to the SAM guidelines as an instruction.) The decided accident management actions are taken by the plant operators, so they have an 'execution' role, but they do not make decisions themselves anymore. The SU Level 2 PSA for Paks includes the dependence between Level 1 PSA and Level 2 PSA actions due to the emergency operating procedure driven transfer to the SAM guidelines. However, other specific types of decision and action, and the associated dependence, need to be modelled in a MUPSA. The emergency response team is responsible for deciding on the use of shared resources between units and introducing restrictive measures at the different units, using input from the shift supervisor. These decisions and actions can impact greatly on the likelihood of, and releases from, multiple accidents. The procedural support for making decisions that affect multiple units and the training of the plant personnel on the treatment of MU accidents are important influences on personnel responses, too. All these aspects need to be factored into the HRA of the MU Level 2 PSA for Paks. This is seen as a very important analysis area that needs much developmental work. The small scale pilot study did not include a full blown HRA. Thus, similarly to the modelling of shared resources, HRA will be a major task in the full scale MUPSA for NPP Paks.

III–5. CURRENT STATUS AND FUTURE PLANS

The feasibility study resulted in the specification of the major technical tasks in a site risk analysis. A small scale pilot analysis on the LOOP IE led to the development and quantification of an initial two-unit Level 1 PSA model for the Paks plant. Based on the findings of this two-stage study, recommendations were made and presented to responsible plant personnel to move forward with the analysis and evaluation of site risk using MU and multisource PSA modelling for a wide range of IEs and POSs. The proposed analysis is now ongoing, with a focus on developing a full scope MU Level 1 PSA for the four units of the Paks NPP to the extent reasonably practicable. The ultimate goal of the follow on analysis is to meet the objectives outlined in Section III–2, to the extent of:

— Quantifying and evaluating Level 1 PSA measures (core damage and fuel damage risk) for the whole site;
— Identifying analysis areas and associated technical issues in need of improvement or refinement to yield credible risk estimates;
— Examining how the Level 1 PSA for the site can be developed into a Level 2 PSA.

The site Level 1 PSA is to be performed in the following main steps:

(1) Identification of IEs important to site risk;

(2)	Modifications to the unit specific PSA models in order to enable the development of a MUPSA model in a common framework;
(3)	Development of an initial MUPSA model based on linking the modified unit specific PSA models;
(4)	Specification and modelling of shared resources that need to be included in the MUPSA model;
(5)	Modelling of human responses to IEs that affect multiple plant units;
(6)	Integration of unit specific PSA models for external events into the MUPSA model;
(7)	Risk quantification.

III–6.	CONCLUSIONS

Site level risk assessment was found to be necessary for NPP Paks to enable an improved characterization of risk and a better understanding of plant vulnerabilities. Based on the conclusions of the feasibility study, the assessment appeared feasible, although the need was identified for substantial further analysis and developmental work. Development and handling of a MUPSA model, consequence modelling for external events (including the treatment of correlated failures), modelling the use of shared resources by multiple plant units and HRA for MU accidents have to be highlighted in this respect. The pilot study led to a refined proposal for the practical steps to be followed during site risk assessment.

A full scope Level 1 PSA for the Paks site is in preparation by making use of the achievements of the preparatory analyses performed so far. The site level risk assessment for Paks will continue in accordance with the work plan developed for the project.

REFERENCES TO ANNEX III

[III–1]	INTERNATIONAL ATOMIC ENERGY AGENCY, Consideration of External Hazards in Probabilistic Safety Assessment for Single Unit and Multi unit Nuclear Power Plants, Safety Reports Series No. 92, IAEA, Vienna (2018).
[III–2]	CANADIAN NUCLEAR SAFETY COMMISSION, Summary Report of the Int. Workshop on Multiunit Probabilistic Safety Assessment, CNSC, Ottawa, ON (2014).

Annex IV

REPUBLIC OF KOREA — EXPERIENCE WITH MUPSA

Following the Fukushima Daiichi accident in 2011, concerns about MU safety have been raised worldwide. Based on the reference data series from the IAEA [IV–1], around 90% of the nuclear sites in the world have two or more reactor units, while some States, such as Canada, France, Japan and China, have nuclear sites containing more than six units. The Republic of Korea also has nuclear sites with six or more reactor units on a single site, and MU safety in particular has been an important issue in the construction permit process for Shin-Kori Units 5 and 6. In this regard, several MUPSA projects have been launched by the related Korean regulatory body, licensee and research institute, each with their own objectives.

This annex describes the status of Korean NPP sites, as well as the main results of previous and current MUPSA projects in the Republic of Korea. The annex includes inputs from KAERI (Section IV–1), KHNP (Section IV–2) and Sejong University (Section IV–3).

IV–1. KAERI MUPSA PROJECT

Immediately after the Fukushima Daiichi accident, the first MUPSA project was planned by the Korean government and performed by KAERI from March 2012 to February 2017. Following this initial project, KAERI started a new government-funded MUPSA research project in March 2017 to address several issues that were not resolved in the previous one. Section IV–1.1 presents the main results of the previous project, and Section IV–1.2 describes the current status of the ongoing project.

IV–1.1. Previous MUPSA research project (2012–2017)

This initial project developed a methodology and software tools for MUPSA, including Levels 1, 2 and 3 [IV–2 to IV–6]. This project was not intended to perform a MUPSA for a specific real NPP site or to identify site specific vulnerabilities and potential for improvements; rather, the main objective of the project was to develop and validate a methodology and software tools for MUPSA. Since all four NPP sites in the Republic of Korea have six or more reactor units, the developed methodology and software tools are intended to be applicable to a MUPSA for NPP sites with six or more units.

To validate the methodology and software tools, MUPSA models were developed and then site risk was assessed for a reference site with six reactor units. A MUPSA model was developed for each of the following IEs:

(1)   MU LOOP;
(2)   MU loss of ultimate heat sink (UHS);
(3)   MU seismic events;
(4)   MU tsunami events;
(5)   Simultaneous occurrence of independent SU initiators (SUIE) in multiple units.

The MUPSA models were subject to the following assumptions:

(a)   The reference NPP site has six identical units (i.e. six optimized power reactor 1000 (OPR-1000) reactors).[1]
(b)   All SSCs except for diesel generators are identical across the units, and their reliability data (including seismic and tsunami fragility) are the same.
(c)   For EDGs and AACDGs, the current status of two Korean NPP sites was considered. That is, Units 1 and 2 have the same type of EDG and share an AACDG, and Units 3 to 6 have the same type of EDG (a type different from those for Units 1 and 2) and share another AACDG.
(d)   Operators are different between units, but the procedures for operating, testing and maintenance are the same, and the HEP for a certain HFE is the same.
(e)   All six units are in operation at full power. Low power and shutdown operation modes are considered only in the MU LOOP model as a pilot study. SFPs are not included in all MUPSA models.
(f)   A MUIE challenges all six units simultaneously, or nearly simultaneously, and its impact on each unit (e.g. peak ground acceleration in seismic PSA) is the same.
(g)   Interunit seismic correlation is the same regardless of the distance between each pair of units.
(h)   The ground levels of all six units are the same (10 m).
(i)   In the case of a simultaneous occurrence of independent SUIE in multiple units, the occurrence of an SUIE in a specific unit (i.e. the initiating unit) does not affect the probability that a subsequent unit or units at the same site

---

[1]   Actually, the reference site has two different types of operating unit: two Framatome-designed units (France CPI) and four OPR-1000 units. In addition, two APR-1400 units are under construction.

will experience an SUIE. The 'simultaneity' of two or more SUIE is defined as the occurrence of those events within 72 hours.

(j) Adverse effects of core damage or radioactive releases from one unit on the other units are not considered.[2]

(k) The new set of mitigation equipment installed as part of the post-Fukushima actions in the Republic of Korea (e.g. portable diesel generators and pumps) is not considered because related reliability data were not yet available at the time of the project.

IV–1.1.1. MUPSA procedure

Figure IV–1 shows the high level procedure that was used for MUPSA in this project. The following briefly describes how each step of the procedure was performed.

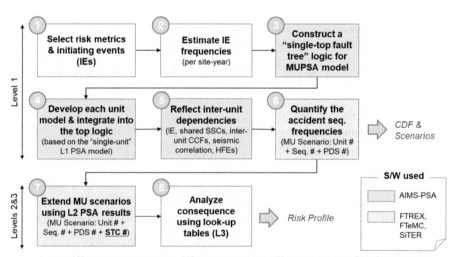

Acronyms: SSC (structure/system/component), CCF (common cause failure), HFE (human failure event), PDS (plant damage state), CDF (core damage frequency), STC (source term category), FTREX (fault tree reliability evaluation expert), FTeMC (fault tree top event probability evaluation using Monte Carlo simulation), SiTER (splitter and integrator for total estimation of site risk)

FIG. IV–1. Procedure of MUPSA.

---

[2] In the subsequent MUPSA project, an approach to modelling and estimating these effects was developed.

Step 1: Select risk metrics and IEs

The first step in the MUPSA procedure is to select site level or MU risk metrics and IEs to be analysed. Examples of risk metrics for MUPSA are site CDF and MUCDF.

As mentioned above, five types of IE were selected: MU LOOP, MU loss of UHS, MU seismic events, MU tsunami events and the simultaneous occurrence of independent SUIE in multiple units. Seismic events were divided into five events of different severity (peak ground acceleration): 0.3g, 0.5g, 0.7g, 0.9g and 1.1g. For each seismic event, a total of 11 seismically induced IEs (e.g. seismically induced LOCA) were considered. Tsunami events were divided into two events of different severity: a tsunami of 5 to <10 m and a tsunami of 10 m or higher.

Step 2: Estimate IE frequencies

The second step is to estimate the frequency of each IE to be analysed. From the MUPSA perspective, it is most convenient to measure IE frequencies on a per site-year basis, rather than on a per reactor-year basis [IV–7].

As in SUPSAs, the frequency of each MUIE can be estimated using different approaches based on its characteristics. For internal events (e.g. MU LOOP), frequency distributions are generally obtained from plant specific or generic industry data. For extremely rare IEs (e.g. MU loss of UHS), frequencies can be estimated by engineering judgement [IV–8]. For external events such as seismic and tsunami events, frequencies are generally based on a site specific probabilistic hazard analysis. An example of applying different approaches to frequency estimation in accordance with the IE category can be found in Ref. [IV–3].

Step 3: Construct a single-top fault tree logic for MUPSA model

The third step is to construct a top logic for the MUPSA model to be developed. As the number of units to be considered increases, the size and complexity of the related MUPSA model also increase. Particularly, when considering three or more units, the event tree method for accident sequence modelling is not applicable. Therefore, the proposed methodology does not develop MU event trees but constructs a single-top fault tree logic and integrates each individual unit model in the form of a single-top fault tree.

Figure IV–2 shows an example single-top fault tree logic for a MUPSA model for a six-unit site. Here, the top event represents core damage in at least one of the six units, which includes all possible combinations of unit core damage results. The fault trees under the top event are distinguished by the number of

167

failed units, and a tag event with a probability of 1.0 for each number of failed units (e.g. '#1UNIT', '#2UNITS') is used in order to prevent the MU cut sets of two or more units failing from being subsumed by SU cut sets. This modelling approach makes it possible to calculate a given risk metric (e.g. site CDF) for the top event, as well as for each number of failed units, **with a single quantification.** More details on this modelling approach can be found in Ref. [IV–6].

Steps 3 to 5 employed AIMS-PSA software [IV–9] in developing the MUPSA model.

Step 4: Develop each individual unit model and integrate into the top logic

The fourth step is to develop each individual unit model and integrate the models into the top logic for the particular MUPSA model. Each individual unit model is based on the related SUPSA model in the form of a single-top fault tree, with modifications made before integration into the top logic as follows:

(a) Fault tree logics unrelated to the IE being analysed were deleted, leaving only the logics related to the IE. This reduces the size of each unit model.
(b) Each individual unit model was constructed so that each accident sequence includes its particular PDS information to make it easier to link with MU Level 2 PSA.

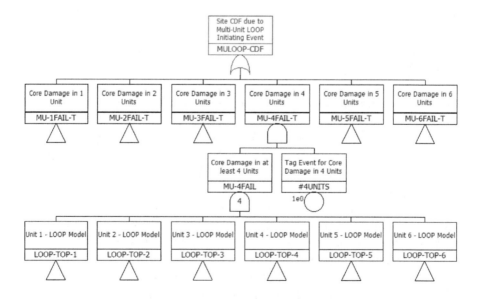

FIG. IV–2. Example single-top fault tree logic for a MUPSA model.

(c) In developing the seismic MUPSA model, the seismic event itself at each seismic interval (e.g. 0.2–0.4g) was modelled as the only IE, with all seismically induced IEs changed to BEs, each with an associated conditional probability of occurrence given the seismic event.

(d) Each unit model was distinguished by changing the names of all gates and BEs. Basically, the name of each gate or BE was prefixed with a number representing its unit: 1, 2, 3, 4, 5 or 6 (e.g. '1GIE-LOOP'). Regarding IEs, the same BE for a specific IE was applied equally to all six unit models without distinguishing between units, because it was assumed that a MUIE challenges all six units simultaneously.

Step 5: Reflect interunit dependencies

The fifth step is to reflect interunit dependencies in the MUPSA model. Interunit dependencies are taken into account for the following aspects:

— MU (or common cause) IEs;
— Shared SSCs between multiple units;
— Dependencies between HFEs in different units;
— Interunit CCFs;
— Interunit seismic correlation.

For a MUIE, the same event name was applied equally to all six unit models without differentiating between units.

In terms of shared SSCs, OPR-1000 plants have a very limited number of SSCs that are shared between multiple units; however, among these, the availability of the AACDG is particularly important from the MUPSA perspective. Since current operating procedures do not dictate unit priority for a shared AACDG, it was assumed that the priority for the two AACDGs is given in the order of Unit 1 → Unit 2 and Unit 3 → Unit 4 → Unit 5 → Unit 6, respectively. In cases where SBO events simultaneously occur in multiple units, the AACDG is available only to the unit with the highest priority (i.e. not available to the other unit(s) with lower priorities) and switching its connection from one unit to another was not credited (see Fig. 3 of this publication for an example fault tree model for a shared AACDG).

Regarding dependencies between HFEs, since each OPR-1000 unit is controlled by its own operating crews in separate MCRs, operator actions performed in the MCRs are regarded as completely independent between units. However, operator actions performed outside the control room, such as off-site power recovery actions, can be considered as dependent. Given the occurrence of a MU LOOP IE, off-site power recovery actions were assumed to be completely dependent between units, and hence the same BE was applied to all six unit models for a given duration (e.g. 7 hours).

Interunit CCFs were considered only for risk significant BEs that have Fussell–Vesely importance values of 0.01 or higher in the base SUPSA model. For each selected component except for the diesel generators, all the components of the six units were grouped into a common cause component group (CCCG) and all possible CCF BEs were modelled. The five diesel generators (including an AACDG) of Units 1 and 2 were grouped into a CCCG, and the nine diesel generators (including an AACDG) of Units 3 to 6 were grouped into another CCCG. For calculating interunit CCF BE probabilities, alpha factors from NUREG/CR-5497 [IV–10] were used, with the mapping up technique [IV–11] employed to obtain impact vectors for large CCCG sizes. However, this approach (i.e. modelling all possible CCF combinations) results in a MUPSA model too large to be quantified in a reasonable time and also makes the estimation of required CCF parameters more complicated. Therefore, in the subsequent MUPSA project, a pragmatic approach to modelling interunit CCFs and estimating their probabilities was developed, which is described in Section 4.4.5.2.

Interunit seismic correlations were considered for two types of events: all seismically induced IEs, and seismically induced fragility BEs with Fussell–Vesely importance values of 0.01 or higher in the base SU seismic PSA model. For each of these events, interunit seismic CCFs (CCCG size = 6) were modelled. Also, interunit seismic correlation was defined at the plant (or unit) level rather than at the individual SSC level. That is, if an interunit seismic correlation of 0.3 was assumed, the interunit correlation of each SSC in the model was regarded as the same. Since it is very difficult to calculate interunit seismic correlation, site CDF was estimated for four different cases, namely, correlations of 0.0, 0.3, 0.7 and 1.0, and the results were compared via sensitivity analysis.

More details on this step are described in Ref. [IV–3].

Step 6: Quantify the accident sequence frequencies

The sixth step is to quantify the accident sequence frequencies. This quantification was basically performed using AIMS-PSA software [IV–9] and the Fault Tree Reliability Evaluation eXpert (FTREX) quantification engine [IV–12]. In addition, the SiTER (splitter and integrator for total estimation of site risk) code was used to find and delete nonsense or duplicate cut sets. The truncation limit for the quantification of each model was determined by lowering it by orders of magnitude until site CDF converged.

Cut set based quantification using rare event approximation or minimal cut upper bound (MCUB) approximation can significantly overestimate site CDF when the model has numerous high probability BEs (e.g. a seismic PSA model).

Therefore, in this case, quantification was performed using fault tree top event probability evaluation using Monte Carlo simulation (FTeMC).

Figure IV–3 shows the software tools developed for MUPSA and their relationships. More details on these tools can be found in Ref. [IV–6].

Level 1 MUPSA ends after Step 6. For Levels 2 and 3 MUPSA, Steps 7 and 8 are necessary.

Step 7: Extend MU scenarios using Level 2 PSA results

In this step, core damage accident scenarios resulting from Level 1 MUPSA are extended to Level 2 scenarios, including information on the source term release category (STC). Here, Level 1 and Level 2 scenarios include both SU and MU accident scenarios.

As described in Step 4 above, each core damage sequence has PDS information. In a SU Level 2 PSA, a containment event tree is developed to analyse accident progression, including consideration of severe accident phenomena, where each PDS is used as the initial condition of the containment event tree analysis. Each PDS is mapped to all containment event tree sequences, and each sequence is mapped to a specific STC. Therefore, for each PDS, the fraction of each STC can be calculated from a SU Level 2 PSA model. Figure IV–4 represents an example of extending Level 1 scenarios (given the occurrence of a MU LOOP IE) to Level 2 scenarios using the PDS-STC fraction table obtained from a base SU Level 2 PSA model with 39 PDSs and 21 STCs. If each unit has a separate SU Level 2 PSA model, a separate PDS-STC fraction table

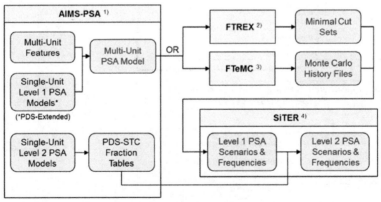

1) AIMS-PSA: PSA Software for Event Tree & Fault Tree Modeling
2) FTREX: Cut Set Generation Engine
3) FTeMC: Fault Tree Analysis Tool using Monte Carlo Approach
4) SiTER: Multi Unit Quantification Tool

FIG. IV–3. Software tools and their relationships.

can be obtained for each unit. However, since it was assumed that all six units are identical in this project, the same table was used for all the units.

The frequency of each extended scenario is calculated by multiplying the frequency of the corresponding Level 1 scenario by the fraction. This can be performed using the SiTER code [IV–6]. Such calculation allows the accident scenarios to be seen in terms of source term release. More details on this step are described in Ref. [IV–4].

Step 8: Analyse the consequences of accident scenarios using look up tables

In this step, an off-site consequence analysis is performed for the extended accident scenarios resulting from Step 7.

One of the major difficulties in Level 3 MUPSA is that the number of combinations of accident scenarios increases exponentially with the number of STCs and the number of units on a site. For example, for 21 STCs and six units, a total of 21^6 (i.e. ~8.58 × 10^8) MU source term release scenarios need to be considered. In such cases, it is impractical to perform consequence analyses for all possible combinations of accident scenarios.

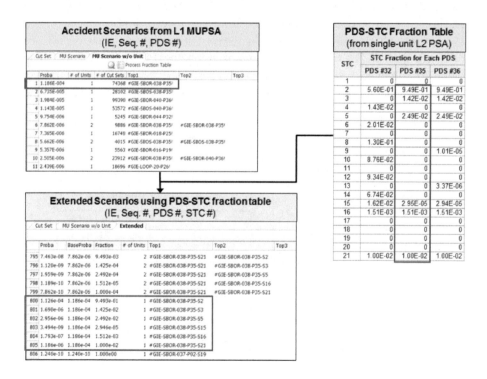

FIG. IV–4. Example of extending Level 1 scenarios to Level 2 scenarios.

Therefore, a look up table approach has been proposed. In this approach, two look up tables need to be constructed in advance: one for early fatalities and the other for latent cancer fatalities. Each source term release from a unit has two parameters: timing of release and magnitude of release (based on ^{137}Cs release fractions). For simplicity, only two timing categories, 'early release' and 'late release', are considered for each STC. The magnitude of release of each STC is converted into a ratio to the greatest magnitude of release (i.e. release of STC-16) for both early and latent health effects, respectively.

Each look up table is two dimensional, with the x-axis representing the sum of the relative magnitudes of early releases from one or more units and the y-axis representing the sum of the relative magnitudes of late releases from one or more units. The value in each table cell represents the conditional consequence (i.e. early fatalities or latent cancer fatalities) given the occurrence of a specific accident scenario, and it is obtained from a consequence analysis for the scenario. The relative magnitude of a release from one unit ranges from 0.0 to 1.0. Therefore, when six units are considered, the sum of the relative magnitudes on each axis ranges from 0.0 to 6.0. In this approach, it is assumed that MU accidents occur in the same place (i.e. only one point).

When the two look up tables are completed, the consequences of each accident scenario resulting from Level 2 MUPSA (Step 7) can be obtained directly from the tables. An example of the MU accident scenario is shown in Table IV–1. For early fatalities, the sum of the relative magnitudes of early releases from the three units is 0.002, and the sum of the relative magnitudes of late releases from the units is 0.490 (0.270 + 0.220). For latent cancer fatalities, the former is 0.001 and the latter is 0.063 (0.003 + 0.060). Therefore, the early fatalities resulting from the scenario correspond to the value at [$x = 0.002$, $y = 0.490$] of the look up table for early fatalities, and the latent cancer fatalities correspond to the value at [$x = 0.001$, $y = 0.063$] of the look up table for latent cancer fatalities. Since the intervals between the rows and columns of the tables are 0.2, values not given in the table can be estimated by interpolation between two adjacent table entries. More details on this step are described in Ref. [IV–5].

IV–1.1.2. Example MUPSA results

Although the scope of this project includes Level 2 and Level 3 MUPSA, this section focuses on Level 1 MUPSA results. The results of Levels 2 and 3 MUPSA can be found in Refs [IV–4, IV–5]. As mentioned above, a Level 1 MUPSA model was developed and site CDF was estimated for each of four representative MUIEs, as well as for the case of the simultaneous occurrence of independent SUIE in two or more units. A reference NPP site with six identical OPR-1000 units was considered, and full scale Level 1 PSA models for a specific

TABLE IV–1. EXAMPLE OF A THREE UNIT SOURCE TERM RELEASE
SCENARIO

Unit	STC released	Timing of release	Magnitude of release	
			For early fatalities	For latent cancer fatalities
Unit 1	STC-2	Early release	0.002	0.001
Unit 2	STC-8	Late release	0.270	0.003
Unit 3	STC-21	Late release	0.220	0.060

OPR-1000 plant were used as the base SU models. The major results of the Level 1 MUPSA can be summarized as follows (see Ref. [IV–3] for further details):

(1) In the case of a six-unit LOOP IE, single-unit CDF contributed 92% of site CDF, two-unit CDF accounted for 7%, and the frequency of core damage in three or more units was negligible (less than 1%). Another important finding was that the separation of LOOP duration curves for SU and MU LOOP events had a considerable impact on site CDF (increased by 70%). In addition, there was not much difference (only ±3%) in the site CDF between the case where only at-power operations were considered and the case where both at-power and shutdown operations were considered. This small difference is because the fraction of time in which each unit is in POS with a relatively high CCDP is quite small, whereas the fraction of time spent in POS with low CCDPs (e.g. maintenance while defuelled) is large.

(2) In the case of a six-unit loss of UHS IE, single-unit CDF accounted for 98% of site CDF, two-unit CDF contributed only 2%, and the frequency of core damage in three or more units was negligible.

(3) For MU seismic events, site CDF was estimated for four cases, assuming interunit seismic correlations (0.0, 0.3, 0.7 and 1.0), with the following trends found:

(i) At a certain seismic hazard interval (e.g. 0.2–0.4g), as the interunit seismic correlation increased, the number of core damage units increased, whereas site CDF itself decreased;

(ii) Given a specific interunit seismic correlation (e.g. 0.3), as the seismic magnitude (peak ground acceleration) increased, the number of core damage units also increased;

(iii) Even when the interunit seismic correlation was 0 (i.e. full independence between units), MUCDF accounted for about 50% of total site CDF, including five intervals (almost 100% at 0.6g or greater);

(iv) The interunit seismic correlation of 0.3 had very little impact on the results compared to those from a correlation of 0, especially at lower peak ground acceleration levels;

(v) The contribution to site CDF from relatively high magnitude seismic intervals (0.6g or greater) increased as the interunit seismic correlation increased.

(4) In the case of MU tsunami events, the number of units experiencing core damage depended strongly on the maximum run-up height of the tsunami event. When the run-up height was lower than the ground level of the site (<10 m), SUCDF contributed 98%. In contrast, when a tsunami event exceeding the ground level occurred, six-unit CDF contributed 97%.

(5) The contribution to site CDF from the simultaneous occurrence of independent SUIE (including internal events, internal floods and internal fires) in two or more units at the reference site was sufficiently low to be neglected.

Although the methodology and software tools were applied to a six-unit NPP site in this project, they are also applicable to MUPSA for a wide range of MU sites, from two-unit sites to NPP sites with seven or even more units.

IV–1.2. New MUPSA research project (2017–2021)

A new project started in March 2017 and is scheduled to end in December 2021. To address a number of issues that were not resolved in the previous project, the following tasks have been performed or are being performed.

IV–1.2.1.   Interunit CCF

A pragmatic approach to modelling interunit CCFs was developed [IV–13]. This approach is intended to be applied to MUPSA involving a large number of NPP units (six or more) and includes methods for modelling CCF combinations and estimating their probabilities.

Regarding the selection of CCF combinations to be modelled, this approach employs separate strategies for intra- and interunit CCFs. For intraunit CCFs, it does not change the combinations that are already included in the SUPSA model for each unit, where it is common to model all possible CCF combinations for major components. Thus, in many cases, all possible intraunit CCF BEs are modelled at the individual 'component' level. For interunit CCFs, though, this

TABLE IV–2. COMPARISON OF CCF COMBINATIONS TO BE
MODELLED BETWEEN CCF MODELLING APPROACHES (THREE
UNITS, TWO TRAINS PER UNIT)

Number of failed components	CCF combinations to be modelled[a]				
	Beta factor model	Two tiered beta factor model	Simplified MGL model	Alpha factor model	Proposed approach
2	—[b]	AB, CD, EF	AB, AC, ..., EF (15)	AB, AC, ..., EF (15)	AB, CD, EF
3	—	—	ABC, ABD, ..., DEF (15)	ABC, ABD, ..., DEF (15)	—
4	—	—	—	ABCD, ..., CDEF (20)	ABCD, ABEF, CDEF
5	—	—	—	ABCDE, ..., BCDEF (6)	—
6	ABCDEF	ABCDEF	ABCDEF	ABCDEF	ABCDEF
Total number of CCF BEs	1	4	31	57	7

[a] Unit 1 has components A and B; Unit 2 has C and D; Unit 3 has E and F. MGL: multiple Greek letter.
[b] not applicable

approach models all possible 'unit combinations' at the unit level, and thus a single interunit CCF BE is modelled for each unit combination, regardless of the number of components in each unit (e.g. intraunit CCCG size). See Fig. 4 in Section 4.4.5 for an example of applying this modelling approach to a three-unit case.

Table IV–2 compares the CCF combinations to be modelled using the proposed approach with the existing approaches: beta factor, two tiered beta factor [IV–14], simplified multiple Greek letter (MGL) [IV–15] and alpha factor models. As expected, the proposed approach significantly reduces the total number of CCF BEs to be modelled with increasing numbers of units or components, compared to the simplified MGL and alpha factor models. Moreover, unlike the beta factor and two tiered beta factor models, this approach does not mask out intermediate CCF combinations (e.g. ABCD, ABEF and CDEF in Table IV–2) by modelling all possible unit combinations.

As for CCF parameter estimation, the proposed approach assumes that interunit CCF events to be modelled (which are determined using the method described above) are subsets of the intraunit complete CCF event, in which all the components in a SU fail. Therefore, the intraunit complete CCF parameter is shared with interunit CCF BEs to be modelled. To this end, the fraction of each interunit CCF combination is evaluated by considering two variables: a component specific parameter (k) reflecting its characteristics and the interunit CCF correlation (r) between the units included in the combination.

The component specific parameter k is defined as the probability that the cause of the intraunit complete CCF would have failed all identical components in an additional identical unit. This parameter ranges from 0 to 1, depending on component specific characteristics, experience data and the number of components in the additional unit (i.e. intraunit CCCG size). If it is difficult to precisely determine the value, $(0.5)^n$ can be used as a default, where n is the intraunit CCCG size.

Interunit CCF correlation r is defined as the degree to which multiple units share the same causal mechanisms of failure for a specific component type. It also ranges from 0 to 1, where 0 indicates completely different components across the units and 1 indicates identical components across the units. The value of r is determined using a decision tree (see Fig. 5 in Section 4.4.5). By considering interunit CCF correlation, the proposed approach makes it possible to deal with 'non-identical but partially correlated' components and their asymmetrical relationships.

The effectiveness of this approach was demonstrated by application to cases with different numbers of identical units (2, 3, 4, 6, 10 and 12 units), as well as to cases where non-identical units are included in MUPSA. A simplified fault tree model was developed for estimating the conditional probability that one or more units will experience SBO given a MU LOOP IE as an example MUPSA model, and for each number of units considered, five MU SBO models were compared, each using a different approach to CCF modelling (i.e. the beta factor, the two tiered beta factor, simplified MGL, the alpha factor and the proposed approach).

The results showed that for all numbers of units, while site SBO probability was not significantly affected by the particular CCF modelling approach used, the MU SBO probabilities differed significantly depending on the approach. As the number of units increased, the beta factor, two tiered beta factor and simplified MGL models provided increasingly conservative estimations of the probability of all units failing and thus yielded underestimated probabilities of 2 to $(n-1)$ units failing. Meanwhile, the alpha factor model exponentially increased the number of CCF combinations to be modelled with increasing number of units, and therefore automatic generation of the CCF fault tree logic using the given PSA software failed in the 10 and 12 unit cases.

On the other hand, when applying the approach proposed in this paper, CCF modelling and fault tree quantification were possible even when 12 units at a site were considered. Furthermore, in the two-, three-, four- and six-unit cases, compared to the alpha factor model (i.e. modelling all possible CCF combinations), the proposed approach estimated the conditional SBO probability for each number of SBO units a little more conservatively (but not significantly so), while still providing similar relative proportions of the number of SBO units. These results indicate that the proposed approach to CCF modelling is effective for MUPSAs, especially when considering NPP sites with a large number of units. Table IV–3 compiles the results of the four-unit case as an example. Here, the SU SBO probability means the probability that only one unit experiences SBO, not including the cases of SBO in two or more units.

IV–1.2.2. POS combinations

An approach to consider numerous combinations of POSs in MUPSA was developed [IV–16]. This approach consists of three main steps. In Step 1, an integrated model for each unit is developed by combining its full power and low power and shutdown (LPSD) PSA models in the form of a single-top fault tree. This integrated SU model needs to include BEs representing the 'fraction of time' spent in each POS, including the full power state, which can be easily calculated based on the full power and LPSD PSA results or operating experience/plan for the unit. Figure IV–5 shows an example integrated SU model that consists of fault tree logics for all POSs. The yellow boxes in this figure are BEs representing a fraction of time. These BEs need to be modelled as mutually exclusive events because each unit can only be in one POS at a time. In addition, one of these events needs to occur for each unit.

In Step 2, the SU models are integrated into the top logic for a MUPSA model, which is also in the form of a single-top fault tree (see Fig. IV–2 for an example).

In Step 3, the accident sequences are quantified. If the size of each integrated SU model or the number of units is large, the integrated MUPSA model can become too complicated to be quantified using available software. In this case, each integrated SU model can be simplified by screening out non-risk significant accident sequences. Then, Steps 2 and 3 are repeated. For the selection of risk significant accident sequences, three levels of screening criteria are applied in

TABLE IV–3. COMPARISON OF QUANTIFICATION RESULTS FOR THE FOUR-UNIT CASE [IV–13]

Number of SBO units	Conditional SBO probability given four-unit LOOP (%)				
	Model 1	Model 2	Model 3	Model 4	Model 5
1	1.34E-02 (97.2%)	1.44E-02 (99.0%)	1.36E-02 (98.2%)	1.37E-02 (98.4%)	1.39E-02 (97.2%)
2	6.73E-05 (0.5%)	7.75E-05 (0.5%)	7.49E-05 (0.5%)	1.87E-04 (1.3%)	3.36E-04 (2.3%)
3	1.50E-07 (0.0%)	1.86E-07 (0.0%)	2.13E-07 (0.0%)	3.09E-05 (0.2%)	6.08E-05 (0.4%)
4	3.13E-04 (2.3%)	6.97E-05 (0.5%)	1.69E-04 (1.2%)	3.01E-06 (0.0%)	5.13E-06 (0.0%)
Sum	1.38E-02	1.45E-02	1.38E-02	1.39E-02	1.43E-02

Note: Models 1 to 5 represent the four-unit SBO models using the beta factor, the two tiered beta factor, simplified MGL, the alpha factor and the proposed approach, respectively.

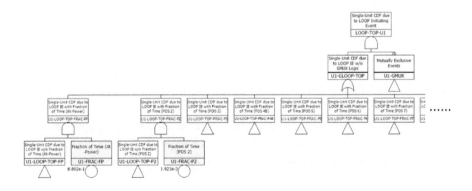

FIG. IV–5. Example integrated SU model combining full power and LPSD PSA models.

stages based on both the individual contribution (percentage) of each accident sequence and the summed contribution of the selected accident sequences.

In this regard, this approach does not select a representative set of POS combinations, but rather focuses on risk significant 'accident sequences' for each unit.

To examine the applicability of this approach, site CDF due to a MU LOOP IE was estimated for three cases with different numbers of units at a site: two-unit, four-unit and six-unit cases. As a result, in the two- and four-unit cases, site CDF due to a MU LOOP was successfully calculated without screening out non-risk significant accident sequences for each unit. On the other hand, in the six-unit case, quantification using FTREX failed without screening. However, it succeeded with screening out the accident sequences with an individual contribution lower than 0.1% in each integrated SU model. In this case, about 80% of the accident sequences in each SU model were screened out, thereby reducing the total number of events by about 40%. In addition, it was found that the resulting MCSs in each case covered a large number of POS combinations, and non-risk significant POS combinations (e.g. four or more units are in shutdown) were truncated by the cut-off value (1×10^{-14}/year).

IV–1.2.3. Dynamics of MU accidents

Practical approaches to considering the dynamics of MU accidents (e.g. cascading effects, timing of release) are currently being developed. For example, an approach has been developed to modelling and estimating the adverse effects of core damage or radioactive releases from one unit on operator actions in adjacent units at the same site [IV–17]. This approach consists of the following three steps:

(1) Selection of risk significant core damage sequences of the preceding unit, which are divided into large and non-large release sequences;
(2) Selection of risk significant HFEs in other adjacent units that can be affected by an accident in the preceding unit;
(3) Development of fault trees for the selected HFEs by applying different HEPs, depending on the types of accident sequences of the preceding unit (large or non-large release) and operator actions in adjacent units (MCR or ex-MCR actions).

Figure IV–6 shows an example fault tree for a selected HFE in an adjacent unit (Unit 2) considering the adverse effects from the preceding unit (Unit 1).

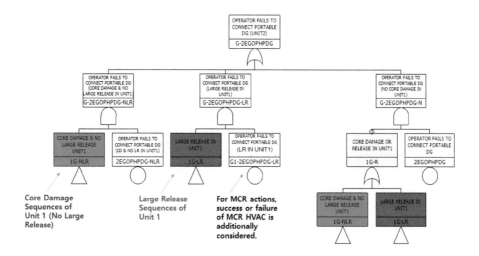

FIG. IV–6. Example fault tree for a HFE of an adjacent unit (Unit 2) considering the adverse effects of an accident in Unit 1.

IV–1.2.4.   MUPSA software tools

MUPSA software tools (see Ref. [IV–6] for more details) are being continuously improved to handle very large fault trees. Also, an integrated software platform for performing MUPSA more effectively and efficiently is being developed.

IV–2.   KHNP MUPSA PROJECT

IV–2.1. Information on the reference site

There are five NPP sites in the Republic of Korea, with each site having more than four reactor units that include various reactor types, including Westinghouse, Framatome, CANDU, OPR-1000 and APR-1400, as shown in Table IV–4.

Although the Saeul site was separated from the Kori site in 2017 in terms of organization, staff, etc., the distance between the two sites is only about 3 km or less. Such a short distance makes it difficult to clearly determine whether the two sites have independent impacts from external hazards such as seismic events, typhoon and so on. Therefore, KHNP considered these two sites as one combined site and selected it as the reference site for its MUPSA project; as a result, the project covers nine reactor sources and ten SFP sources [IV–18].

TABLE IV–4. STATUS OF NPPS IN REPUBLIC OF KOREA

Site	In operation	Under construction	Permanent shutdown	Type of reactor
Kori	5	—[a]	1 (Westinghouse type)	Westinghouse type (PWR): Kori Units 1–4 OPR-1000 (PWR): Shin-Kori Units 1–2
Saeul	2	2 (APR-1400)	—	APR-1400 (PWR): Shin-Kori Units 3–6
Hanul	6	2 (APR-1400)	—	Framatome type (PWR): Hanul Units 1–2 OPR-1000 (PWR): Hanul Units 3–6 APR-1400 (PWR): Shin-Hanul Units 1–2
Hanbit	6	—	—	Westinghouse type (PWR): Hanbit Units 1–2 OPR-1000 (PWR): Hanbit Units 3–6
Wolsong	6	—	—	CANDU type (PHWR): Wolsong Units 1–4 OPR-1000 (PWR): Shin-Wolsong Units 1–2

[a] Data not available.

All the various types of NPPs in the Republic of Korea do not share the same SSCs related to safety functions, based on a Korean regulation requirement, examples of which would be Requirement 33 on the sharing of safety systems between multiple units of a NPP in IAEA Safety Standards Series No. SSR-2/1 (Rev. 1) [IV–19] and 10 CFR Appendix A to Part 50 GDC Criterion 5 from the NRC. Accordingly, KHNP employs only a few shared SSCs, for which related interunit dependencies need to be considered in MUPSA models. Examples include AACDGs and off-site power sources, including switchyards.

IV–2.2. Approach to developing MUPSA models

IV–2.2.1. Scope of MUPSA

KHNP has previously developed SU Level 1 PSA models for all operating modes and SU Level 2 PSA models for full power operation for all units. SU Level 3 PSA for full power operation, however, has been performed only for APR-1400 units. Considering the SUPSA models of the reference site, KHNP has determined that the scope of MUPSA will cover Level 1 and Level 2 PSA for all operating modes.

IV–2.2.2. IE analysis

MUIEs can be defined as events that cause reactor trips in multiple units within a certain period. In general, three kinds of events can be considered as MUIEs [IV–2]:

(1)   Simultaneous occurrence of independent IEs in two or more units;
(2)   Affected MUIEs;
(3)   Common IEs.

As for the first category, the MU risk due to internal events/flooding/fires occurring independently at two or more units was identified to be negligible, and the second category is considered to have little effect because each unit has its own independent structures with very few shared systems or components, according to previous research [IV–3]. Accordingly, KHNP focused on the third category, in which an IE simultaneously causes reactor trips in two or more units at a site.

Considering the scope of SUPSA models, seismic hazards were first determined as MU initiators, which could have concurrent impacts on multiple units at a site. To identify other MU initiators, historical operating experience was reviewed for cases with concurrent reactor trips in multiple units. As a result, MU initiators including LOOP, general transient and loss of circulating water due to typhoon, heavy snow, lightning, marine lives, etc. were identified. Considering the large uncertainties inherent in the hazard analyses of typhoons, heavy snow, lightning, etc., the frequencies of these MU initiators were estimated based on operating experience. Other external hazards, except for seismic hazards and high/low levels of seawater, were qualitatively screened out. Therefore, KHNP has been performing a tsunami hazard analysis in order to consider all possible external hazards in the project. Table IV–5 summarizes the scope of developing the preliminary MUPSA models [IV–20].

TABLE IV–5. SCOPE OF DEVELOPING MUPSA MODELS

Operating mode	MU initiators	IE frequency estimation method	Scope of PSA
All operating modes	LOOP (due to typhoon, heavy snow)	Operating experience	Level 1 and Level 2
	General transient (due to typhoon, lightning, shared systems)		
	Loss of circulating water (due to marine life)		
	Seismic event	Hazard analysis	

IV–2.2.3. Concept of SOSs

In KHNP's MUPSA project [IV–18], a total of nine reactor units are considered. Note that, for example, for about 15 POSs in a SU LPSD PSA, more than 60 billion possible combinations need to be considered, for which model development or management is not possible. Further, even in the case of applying all possible combinations of POSs to MUPSA models, significant insight with respect to MU risk is hardly expected to result. KHNP has therefore introduced the concept of SOSs [IV–21] and has suggested a simple and conservative approach, as follows:

(a) Representative SOSs are defined based on historical experiences of overhaul and the long term plan for overhaul schedules, followed by an estimation of the fraction of time spent in each SOS:

(i) KHNP identified that all nine units would be in full power operation for about 40% of a year. For around 50% of a year, one of the nine units would be in the LPSD operating mode, while the other eight units would be in full power operation. The duration of two units in LPSD corresponds to about 10% of a year. Although KHNP also identified the period of three reactor units in LPSD, this could be negligible. As a result, KHNP considered three different SOSs in MUPSA model development, namely SOS1, SOS2 and SOS3, as shown in Table IV–6.

(ii) Since the duration of the POS with no fuel in the reactor takes up about 43% of the entire overhaul duration of a SU, two additional SOSs (SOS4 and SOS5) were considered in MUPSA model development, as shown in Table IV–6.

(b) Representative units are defined based on engineering judgement, following some assumptions:

(i) As the representative overhaul unit, KHNP designated an old design unit at the Kori site (K3) when considering only one overhaul unit. As the representative SOS unit, which considers two overhaul units with seven full power operating units, the first operating unit of the Saeul site (S3) was designated.

Even though a given unit could be in various POSs during overhaul, it was assumed that the unit is in only one POS, which has the most conservative CCDP. If this assumption produces overly conservative results, more POSs can be considered.

IV–2.2.4.   Modelling structures of MUPSA models

KHNP developed MUPSA models based on SUPSA models, in which typical PSA methods of combining event trees and fault trees are used. Since the size of the MUPSA models, however, is much larger than that of SUPSA models, MUPSA models were developed using the 'single-top fault tree' method,

TABLE IV–6. SOS AND ESTIMATED FRACTIONS OF TIME

SOS	Description	Time fraction of a year
SOS 1	All nine units are in full power operation	44.8%
SOS 2	Eight units are in full power operation (All fuels of one unit (K3) are stored in the SFP)	23.2%
SOS 3	Seven units are in full power operation (All fuels of two units (K3, S3) are stored in the SFP)	4.7%
SOS 4	Eight units are in full power operation and one unit (K3) is in overhaul	22.7%
SOS 5	Seven units are in full power operation and two units (K3, S3) are in overhaul	4.6%

considering the quantification capacity of PSA software [IV–18]. At first, the SUPSA models of each unit were modified into a single-top fault tree as shown in Fig. IV–7. Then, KHNP developed a logic tree to integrate all SUPSA models to consider all possible combinations of accident scenarios in two or more units. In accordance with the different SOSs, MUPSA models were separately developed. Figure IV–8 shows an example structure of MUPSA models for SOS1, and Fig. IV–9 shows MUPSA models for SOS2, to which LPSD SUPSA models were applied for K3.

IV–2.2.5.   Approaches to interunit dependencies

To consider interunit CCF, KHNP selected the important components for which intraunit CCF BEs have Fussell–Vesely importance measures over 0.005. KHNP added interunit CCF BEs to all failure modes of the selected components. A simple modelling approach was applied because the number of interunit CCF events increases as the size of the CCCG becomes larger than that of intraunit CCF. So KHNP modelled one BE of interunit CCF between twin units, among units of the same reactor type and among all units at the site, as shown in Fig. IV–10. A conservative assumption was considered to estimate the related probabilities: 50%, 25% and 10% for intraunit CCF BEs for CCFs between twin units, among units of the same reactor type and among all units at the site, respectively [IV–20].

In seismic PSA, some technical considerations need to be considered in addition to seismic correlation. As for seismic hazard analysis, the same seismic hazard curves and the same fragility capacity for LOOP were applied to all units. Although the units are operating at the same site, ground response or equipment response in individual units against seismic hazards would differ. The effects from these differences, however, were ignored. In this project, MU seismic PSA models were developed for five seismic hazard groups based on the seismic hazard magnitude in order to derive risk insights from different seismic levels. Table IV–7 shows the five groups, along with multiplication factors to the HEP of internal event PSA models, which were simply assumed following engineering judgement. As for seismic correlation, there is no mature method internationally except for using fully correlated (1.0) or fully independent (0.0) correlation. Therefore, KHNP conservatively assumed 1.0 for the identical components between twin units; for example, all component cooling water (CCW) pumps in twin units were fully correlated against seismic hazard at the site. Otherwise, 0.0 was assumed [IV–18].

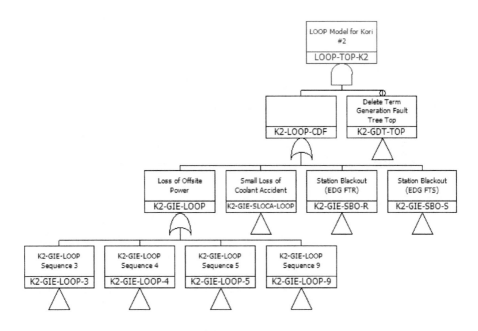

FIG. IV–7. Structure of modified SUPSA models for a MU LOOP event.

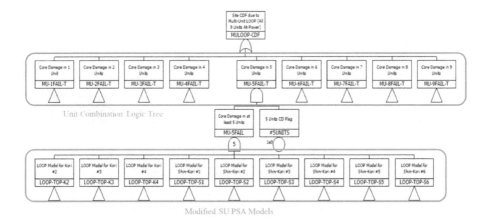

FIG. IV–8. Structure of integrated MUPSA models for SOS1.

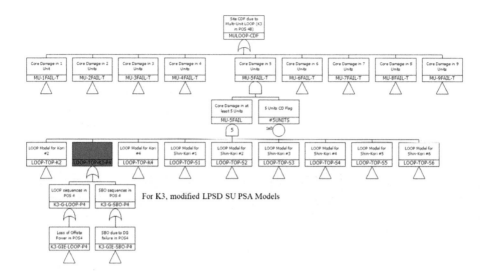

FIG. IV–9. Structure of integrated MUPSA models for SOS2.

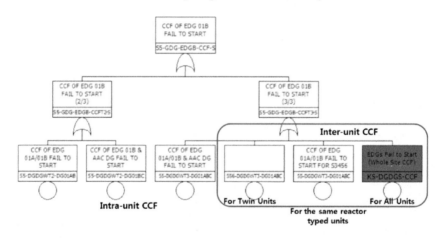

FIG. IV–10. Modelling approach to interunit CCF.

IV–2.3. Risk metrics and results

As there is no performance goal or safety goal for MUPSA, the risk metric of Level 1 MUPSA was determined as MUCDF based on a SUPSA background. Therefore, the CDFs from any two-unit accident sequences to all nine unit accident sequences were estimated. Since issues of MU risk have recently arisen following the construction of new units in the Republic of Korea, KHNP evaluated the results for two cases: one for the existing seven units and the other for nine units including the two newly built units, to identify the increasing level of MU risk due to the addition of the new units at the reference site [IV–20].

Figure IV–11 shows the results of two cases from MU LOOP models. It was identified that the total MUCDFs for both cases corresponded to about 1×10^{-6}/year and that MUCDFs did not significantly increase following the addition of the newly built units to the reference site. In addition, it was found that MUCDF could be reduced to about 50–60% by performing sensitivity analyses on the interunit CCF assumptions and the SOS simplification. As for MU loss of circulating water and MU general transient, MUCDFs were estimated to be about 2×10^{-9}/year and 2×10^{-10}/year, respectively. Based on these results, MU risk could be ignored, and thus it was determined that further analysis for these two initiators would not be performed, such as Level 2 PSA [IV–20].

MUCDFs for seismic hazards were not sensitive to interunit CCF assumptions and the SOS simplification but were sensitive to the magnitude of seismic hazard; MUCDFs for G03–G05 took up about 94% of total MUCDF. The impact of seismic correlation on MUCDF was also identified. For the seismic hazard groups of G2 and G3, the lower the correlation factor, the lower the estimated MUCDF. However, the seismic correlation factor had little impact on MUCDF for the other seismic hazard groups. Figure IV–12 shows the rate of decrease of MUCDF from a sensitivity analysis on the seismic correlation factors.

As for ongoing activities, KHNP has been developing the final MUPSA methodologies, in particular technical considerations such as interunit CCF, seismic correlation, HRA, etc., and performing a quantitative tsunami hazard analysis. KHNP finalized Level 1 and Level 2 MUPSA models including other sources of SFPs in 2021.

TABLE IV–7. SEISMIC HAZARD GROUP AND HEP FOR MUPSA

Seismic hazard group	Magnitude range of seismic hazard	HEP in MUPSA for seismic event
G01	0.1–0.2g	$3 \times \text{HEP}_{\text{Internal event}}$
G02	0.2–0.3g	
G03	0.3–0.5g	$5 \times \text{HEP}_{\text{Internal event}}$
G04	0.5–0.7g	
G05	0.7–1.0g	$10 \times \text{HEP}_{\text{Internal event}}$

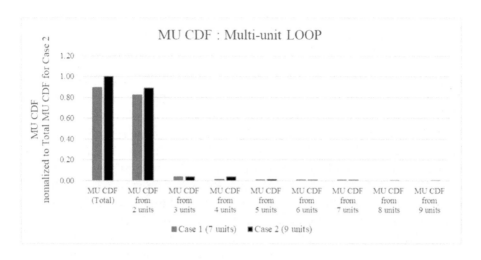

FIG. IV–11. Results of Level 1 MUPSA for MU LOOP (two cases).

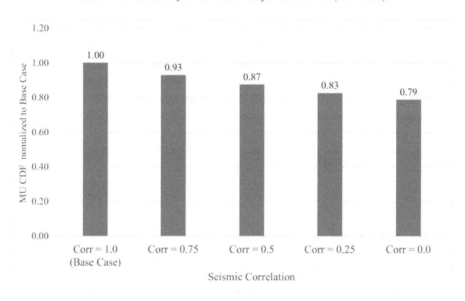

FIG. IV–12. Sensitivity analysis of seismic correlation factors on MUCDF.

IV–2.4. Safety goals

For the safety goals on unit based risk, the Nuclear Safety and Security Committee (NSSC; the nuclear regulatory authority in the Republic of Korea) and the Korea Institute of Nuclear Safety developed the rules, regulatory standards and guidelines for NSSC as a subsequent process of an amendment of the Nuclear Safety Act in 2015. In accordance with the NSSC rules, the following two safety goals are applied as risk target values that the PSA needs to satisfy:

(1) The risk to an average individual in the vicinity of a NPP of prompt fatalities resulting from reactor accidents is not allowed to exceed 0.1% of the sum of prompt fatality risks resulting from all other accidents. The risk to the population in the area near a NPP of cancer fatalities resulting from NPP operation is not allowed to exceed 0.1% of the sum of cancer fatality risks resulting from all other causes, or the equivalent performance goals for the prompt fatality risk and the cancer fatality risk have to be satisfied.

(2) The sum of frequencies for the accident scenarios in which the amount of ^{137}Cs release exceeds 100 TBq has to be lower than 1.0×10^{-6}/year.

The first criterion is adopted from the quantitative safety goals in the severe accident policy, which was announced in 2001.

The equivalent performance goals for prompt fatality risk and cancer fatality risk in the NSSC rules are defined in the Korea Institute of Nuclear Safety regulatory standards/guidelines as follows:

(1) CDF — performance goal equivalent to cancer fatality risk:
    (i) Less than 1.0×10^{-4}/year for operating NPPs;
    (ii) Less than 1.0×10^{-5}/year for new NPPs (e.g. APR-1400 and follow up designs).

(2) LERF — performance goal equivalent to prompt fatality risk:
    (i) Less than 1.0×10^{-5}/year for operating NPPs;
    (ii) Less than 1.0×10^{-6}/year for new NPPs (e.g. APR-1400 and follow up designs).

For the safety goals on site (or MU) based risk, the NSSC launched a new project in 2017 which was completed in 2021. The project aims at developing the regulatory framework for site risk, including site level safety goals and related regulatory standards and guidelines. Since this new project, the NSSC and the Korea Institute of Nuclear Safety are now making efforts for developing a site level risk regulatory framework with other technical support organizations.

IV–3. MU RISK RESEARCH GROUP PROJECT

IV–3.1. Background and definitions

IV–3.1.1.   Background

After the Fukushima Daiichi accident, MU risk became a hot issue in the area of nuclear safety, in particular for the regulatory bodies, because more than six units are located at all nuclear sites in the Republic of Korea. The Kori site (including the Saeul site), in particular, has seven units in operation with another two units under construction. Although there is a trend towards addressing risks in the MU context, the standard methodology is currently under development. Hence, a Multi-Unit Risk Research Group was organized for the R&D project 'Development of Multi-unit PSA (MUPSA) regulation verification technology (2017–2022)' that is supported by the NSSC through the Korea Foundation of Nuclear Safety. Many Korean universities, including Hanyang and Sejong universities and two national institutes, are participating in this project to assess Kori site risk and develop a site risk model.

The Venn diagram for MUPSA CDFs is negated in MUPSA, and DTAs for solving negation in MUPSA are explained for easier understanding of quantification uncertainties in internal and seismic MUPSAs.

IV–3.1.2.   Definitions of site and MUCDFs

As shown in Fig. IV–13, MUPSA for three nuclear units has seven core damage states. The site and MUCDFs for N nuclear units are defined as follows:

$$\text{Site CDF} = p(U1+U2+U3) = p(2^N - 1 \text{ states}) \tag{IV–1}$$

$$\text{Multi-unit CDF} = p(U1U2+U1U3+U2U3) = p(2^N - 1 - N \text{ states}) \tag{IV–2}$$

$$\text{Site CD} = U1+U2+U3 = \begin{cases} U1/U2/U3+ \\ /U1\ U2/U3+ \\ /U1/U2\ U3+ \\ U1\ U2/U3+ \\ U1/U2\ U3+ \\ U1\ U2/U3+ \\ U1\ U2\ U3 \end{cases} \tag{IV–3}$$

$$\text{Multi-unit CD} = U1U2 + U1U3 + U1U3 = \begin{cases} U1\ U2\ /U3\ + \\ U1\ /U2\ U3\ + \\ /U1\ U2\ U3\ + \\ U1\ U2\ U3 \end{cases} \tag{IV-4}$$

Since each core damage state is disjoint,

$$\text{Site CDF} = p(U1 + U2 + U3) = \begin{cases} p(\ U1\ /U2\ /U3) + \\ p(/U1\ U2\ /U3) + \\ p(/U1\ /U2\ U3) + \\ p(\ U1\ U2\ /U3) + \\ p(\ U1\ /U2\ U3) + \\ p(\ U1\ U2\ /U3) + \\ p(\ U1\ U2\ U3) \end{cases} \tag{IV-5}$$

$$\text{Multi-unit CD} = U1U2 + U1U3 + U1U3 = \begin{cases} p(\ U1\ U2\ /U3) + \\ p(\ U1\ /U2\ U3) + \\ p(/U1\ U2\ U3) + \\ p(\ U1\ U2\ U3) \end{cases} \tag{IV-6}$$

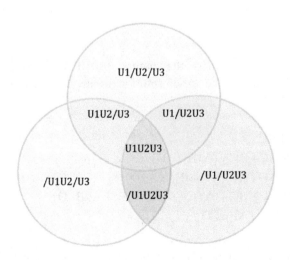

FIG. IV–13. Venn diagram for MUPSA.

IV–3.1.3.   Inclusion–exclusion principles

If a fault tree S has MCS c_i in Eq. (IV–7), the exact probability $P(S)$ can be calculated by inclusion–exclusion probability P_{IE} in Eq. (IV–8). Furthermore, an approximate probability of $P(S)$ can be calculated by means of the MCUB probability P_{MCUB} in Eq. (IV–8). If there are duplicated component failures among MCSs (c_1, c_2, c_3, …), P_{MCUB} is an approximate probability. In contrast, if no MCSs share any component failures, P_{IE} and P_{MCUB} are identical, since $P(c_ic_j)$ is equal to $P(c_i)P(c_j)$. Thus, P_{MCUB} is an exact probability when there are no duplicated component failures in MCSs. This means that the MCUB probability can be either an exact or approximate probability, depending on the MCS structure.

$$S = \bigcup_i c_i = c_1 + c_2 + c_3 \ldots \tag{IV–7}$$

$$\begin{aligned} P(S) &= P_{\mathrm{IE}}(S) \\ &= \sum_i P(c_i) - \sum_{1 \le i < j \le n} P(c_i c_j) + \sum_{1 \le i < j < k \le n} P(c_i c_j c_k) - \ldots \end{aligned} \tag{IV–8}$$

$$\begin{aligned} P(S) &\le P_{\mathrm{MCUB}}(S) \\ &= 1 - \prod_i \left(1 - P(c_i)\right) \\ &= \sum_i P(c_i) - \sum_{1 \le i < j \le n} P(c_i)P(c_j) + \sum_{1 \le i < j < k \le n} P(c_i)P(c_j)P(c_k) - \ldots \end{aligned} \tag{IV–9}$$

IV–3.1.4.   DTA

In order to understand the nature of MUPSA quantification, the DTA is explained here. The exact solution of the Boolean equation can be illustrated as follows:

$$\begin{aligned} &\mathrm{U1/U2/U3} \\ &= (\mathrm{ABD + AC + BC}) / (\mathrm{AB}) / (\mathrm{CD}) \\ &= (\mathrm{ABD + AC + BC})(/\mathrm{A} + /\mathrm{B})(/\mathrm{C} + /\mathrm{D}) \\ &= (\mathrm{ABD + AC + BC})(/\mathrm{A}/\mathrm{C} + /\mathrm{A}/\mathrm{D} + /\mathrm{B}/\mathrm{C} + /\mathrm{B}/\mathrm{D}) \\ &= \mathrm{A}/\mathrm{BC}/\mathrm{D} + /\mathrm{ABC}/\mathrm{D} \end{aligned} \tag{IV–10}$$

The same logic can be solved by the DTA. All negations in a SUPSA and MUPSA are solved by this DTA.

$$U1 / U2 / U3$$
$$= (ABD + AC + BC) / (AB) / (CD) \qquad \text{(IV–11)}$$
$$\approx AC + BC$$

As listed in Table IV–8, the DTA for solving the negation in Eq. (IV–6) can cause extremely overestimated CDFs if event failures are non-rare events. If event failures are rare events, such as, $p(A) = p(B) = p(C) = p(D) = 0.001$, all the successful event probabilities are very close to the probability 1, such as $p(/A) = p(/B) = p(/C) = p(/D) \approx 1.0$, and finally the two probabilities in Eqs (IV–5) and (IV–6) are very similar.

However, if event failures are non-rare events, such as $p(A) = p(B) = p(C) = p(D) = 0.5$, all the successful event probabilities are not close to the probability 1, such as $p(A) = p(B) = p(C) = p(D) = 0.5$, and finally the probability in Eq. (IV–6) is much bigger than that in Eq. (IV–5).

Consequently, the modelling of negations and the DTA application cannot be allowed in a seismic MUPSA that has non-rare events.

Seismic CCDP can be calculated by setting the seismic initiator as TRUE event in Eqs (IV–5) and (IV–6). If DTA is applied to the seismic MUPSA model that has non-rare seismic events, site CCDP and MU CCDP can have a probability much larger than one. So, it can be concluded that the application of DTA has to be avoided for seismic MUPSA.

TABLE IV–8. STATUS OF NPP IN REPUBLIC OF KOREA

Method	Solution for U1/U2/U3	$P(U1 / U2 / U3)$		
		0.001 [a]	0.500 [a]	0.999 [a]
Exact solution	A/BC/D+ /ABC/D	1.996E-06 [b]	1.250E-01 [b]	1.996E-06 [b]
DTA solution	AC+BC	1.999E-06 [c]	3.750E-01 [c]	9.990E-01 [c]
DTA solution	AC+BC	2.000E-06 [d]	4.375E-01 [d]	1.000E+00 [d]

[a] $p(A) = p(B) = p(C) = p(D)$
[b] P_{IE}(A/BC/D+/ABC/D)
[c] $P_{IE} (AB + BC) \, P_{IE}(AB + BC)$
[d] P_{MCUB}(AB + BC)

IV–3.2. Quantification uncertainties in internal events MUPSA

Fault tree analysis is extensively and successfully applied to the risk assessment of safety-critical systems such as nuclear, chemical and aerospace systems. Fault tree analysis had been used together with event tree analysis in the PSA of NPPs. In a PSA, cut sets or MCSs for accident sequences are generated from a set of fault trees and event trees. Each cut set represents an accident sequence that might result in the undesired condition such as core damage. An accident sequence represents successive failures of components or systems after an IE.

Uncertainty in the SUPSA of NPPs can be classified into (1) parameter uncertainty, (2) model uncertainty, (3) completeness uncertainty and (4) quantification uncertainty. Uncertainty sources in PSA are classified into the first three groups [IV–22, IV–23].

Most of the fault tree analysis methods and software for a PSA are based on the cut set based algorithm. They generate cut sets from a fault tree by using the traditional Boolean algebra or zero suppressed binary decision diagram algorithm [IV–24, IV–25] and calculate the top event probability from the cut sets. Although cut set based fault tree analysis has played an important role in SUPSA for NPPs, it is a much more complex and time consuming activity to calculate cut sets in the MUPSA model than in the SUPSA model, as well as perform cut set post-processing, calculate the top event probability from the processed cut sets and calculate importance measures from the processed cut sets. Since the MUPSA fault trees are huge in size and the number of cut sets grows exponentially with the size of a fault tree, cut set based fault tree solvers employ approximations in order to overcome high memory requirements and a long computing time.

In the PSA industry, cut sets can be produced and manipulated by a number of PSA tools. Each PSA tool employs one of the dedicated cut set generation algorithms. One of them is FTREX [IV–12, IV–24], which is the most popular fault tree solver in the USA. It had to be improved to solve MUPSA models with a reasonably low truncation limit, such as 1×10^{-12}.

The scheme in Fig. IV–14 illustrates the MUPSA modelling and quantification stages, the sources of quantification uncertainty in each stage and the efforts to overcome or reduce the quantification uncertainty.

As shown in Fig. IV–14, the sources of quantification uncertainty in MUPSA are:

(1)   Negations in MU level such as U1/U2/U3;
(2)   Cut set truncation that is designed to minimize huge memory requirements;
(3)   DTA for solving accident sequence level negations;
(4)   DTA for solving MU level negations;

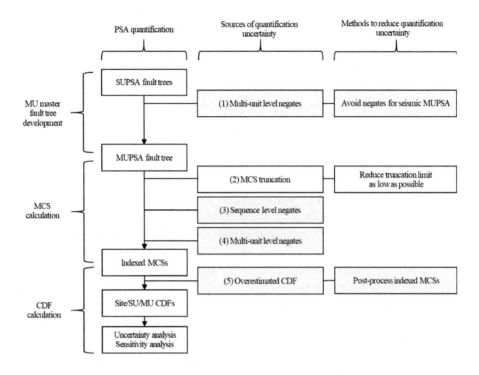

FIG. IV–14. Quantification uncertainty for internal MUPSA.

(5)   Overestimated MCUB probability.

In order to minimize the quantification uncertainty, it is necessary to:

(a)   Correctly compose top level logic in MUPSA (see Section 4.1.2);
(b)   Use a reasonably low truncation limit, such as 1×10^{-12} or 1×10^{-13};
(c)   Carefully trace non-rare events in accident sequences if they cause overestimation of CDF after DTA application;
(d)   Carefully trace non-rare events in the MUPSA model if they cause overestimation of CDF after DTA application;
(e)   Convert cut sets into a binary decision diagram structure [IV–22] to calculate accurate CDF from the final cut sets.

Here, a good cut set generation algorithm is necessary to solve the MUPSA model and generate cut sets with a reasonably low truncation limit. If it is not available, the existing cut set generation algorithm has to be drastically revised. It is noted that the binary decision diagram application to the cut sets cannot overcome the uncertainty resulting from DTA approximation.

IV–3.3. Quantification uncertainties in seismic MUPSA

As shown in Fig. IV–15, there are six sources of quantification uncertainties in seismic MUPSA:

(1) Negations in MU level such as U1/U2/U3;
(2) Seismic failure correlation among NPPs;
(3) Cut set truncation that is designed to minimize huge memory requirements;
(4) DTA for solving accident sequence level negations;
(5) DTA for solving MU level negations;
(6) Overestimated MCUB probability.

In order to minimize the quantification uncertainties, it is necessary to

(a) Correctly compose top level logic in MUPSA (see Eq. (IV–12));
(b) Convert correlated seismic failures into seismic CCFs;
(c) Use a reasonably low truncation limit such as 1×10^{-12} or 1×10^{-13};
(d) Avoid negations in accident sequences (see Eq. (IV–12));

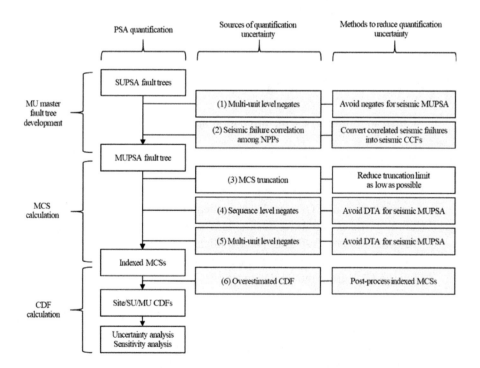

FIG. IV–15. Quantification uncertainty for seismic MUPSA.

(e) Avoid negations in the MUPSA model (see Eq. (IV–12));

(f) Convert cut sets into a binary decision diagram structure [IV–22] to calculate accurate CDF from final cut sets.

In order to minimize overestimated CDFs, negate modelling has to be avoided. As shown in Eq. (IV–12), p(U1/U2/U3) can be calculated without modelling negations U1/U2/U3 in the MUPSA fault tree and DTA for solving U1/U2/U3.

$$
\begin{aligned}
&p\left(\text{U1}/\text{U2}/\text{U3}\right) \\
&= p\left(\text{U1}/\left(\text{U2}+\text{U3}\right)\right) \\
&= p\left(\text{U1}\right) - p\left(\text{U1}\left(\text{U2}+\text{U3}\right)\right) \\
&= p\left(\text{U1}\right) - p\left(\text{U1U2}+\text{U1U3}\right) \\
&= p\left(\text{U1}\right) - p\left(\text{U1U2}\right) - p\left(\text{U1U3}\right) + p\left(\text{U1U2U3}\right)
\end{aligned}
\qquad (\text{IV–12})
$$

The dependency among seismic failures is not explicitly modelled. This dependency is separately assigned with numbers ranging from zero to unity to reflect the mutual correlation level among seismic failures. The determination and calculation of the mutual correlation level among seismic failures is an important issue in a seismic PSA.

The correlated seismic failures of identical components exist in a single MCS in a SUPSA. Usually, if these components are on the same level of the same building, full correlation is assumed and the component failures in the fault tree are replaced by the representative component as AB-A. It guarantees conservative CDF, since $p(AB) < p(A)$.

However, identical components exist across MUs, and the correlated seismic failures of identical components exist across MCSs in a MUPSA. If full correlation is assumed, and the component failures in the fault tree are replaced in the representative component as $A + B \rightarrow A$, it does not guarantee conservative CDF since $p(A + B) > p(A)$.

There has been a great need for development to explicitly model seismic correlation with seismic CCFs. If the correlated seismic failures can be converted into seismic CCFs, an accurate value of a top event probability or frequency of a complex seismic fault tree can be obtained.

The study in Ref. [IV–25] develops and expands a methodology to explicitly model dependency among seismic failures by converting correlated seismic failures into seismic CCFs. The multivariate normal integration for the AND combination probability of correlated seismic failures is:

$$P_{123}(a) = \int_{-\infty}^{\ln\frac{a}{A_{1m}}} \int_{-\infty}^{\ln\frac{a}{A_{2m}}} \int_{-\infty}^{\ln\frac{a}{A_{nm}}} \frac{1}{\sqrt{|\Sigma|(2\pi)^n}}$$

$$\exp\left(-\frac{1}{2} x^t \Sigma^{-1} x\right) dX_1 dX_2 dX_3 \tag{IV-13}$$

$$x^t = \begin{bmatrix} X_1 & X_2 & X_3 \end{bmatrix}$$

$$\Sigma = \begin{bmatrix} \beta_1^2 & \beta_{12}^2 & \beta_{13}^2 \\ \beta_{21}^2 & \beta_2^2 & \beta_{23}^2 \\ \beta_{31}^2 & \beta_{32}^2 & \beta_3^2 \end{bmatrix}, \beta_{ij}^2 = \beta_{ji}^2 = \text{cov}\left(X_i, X_j\right) \tag{IV-14}$$

In this case, corresponding Boolean equation sets using seismic CCFs are given in Eq. (IV–15). Here, three correlated seismic failures are split into seven independent seismic CCFs:

$$X_1 = C_1 + C_{12} + C_{13} + C_{123}$$
$$X_2 = C_2 + C_{12} + C_{23} + C_{123}$$
$$X_3 = C_3 + C_{13} + C_{23} + C_{123}$$
$$X_{12} = C_1 C_2 + C_{12} + C_{123} + C_1 C_{23} + C_2 C_{13} + C_{13} C_{23}$$
$$X_{13} = C_1 C_3 + C_{13} + C_{123} + C_1 C_{23} + C_3 C_{12} + C_{12} C_{23} \tag{IV-15}$$
$$X_{23} = C_2 C_3 + C_{23} + C_{123} + C_2 C_{13} + C_3 C_{12} + C_{12} C_{13}$$
$$X_{123} = C_1 C_2 C_3 + C_{123} + C_1 C_{23} + C_2 C_{13} + C_3 C_{12} + C_{12} C_{13} + C_{13} C_{23} + C_{12} C_{23}$$

The MCUB probabilities of Boolean equations on the right hand sides in Eq. (IV–15) are given in Eq. (IV–16):

$$P_1 = 1 - (1 - Q_1)(1 - Q_{12})(1 - Q_{13})(1 - Q_{123})$$
$$P_2 = 1 - (1 - Q_2)(1 - Q_{12})(1 - Q_{23})(1 - Q_{123})$$
$$P_3 = 1 - (1 - Q_3)(1 - Q_{13})(1 - Q_{23})(1 - Q_{123})$$
$$P_{12} = 1 - (1 - Q_1 Q_2)(1 - Q_{12})(1 - Q_{123})(1 - Q_1 Q_{23})(1 - Q_2 Q_{13})(1 - Q_{13} Q_{23})$$
$$P_{13} = 1 - (1 - Q_1 Q_3)(1 - Q_{13})(1 - Q_{123})(1 - Q_1 Q_{23})(1 - Q_3 Q_{12})(1 - Q_{12} Q_{23}) \tag{IV-16}$$
$$P_{23} = 1 - (1 - Q_2 Q_3)(1 - Q_{23})(1 - Q_{123})(1 - Q_2 Q_{13})(1 - Q_3 Q_{12})(1 - Q_{12} Q_{13})$$
$$P_{123} = 1 - (1 - Q_1 Q_2 Q_3)(1 - Q_{123})(1 - Q_1 Q_{23})(1 - Q_2 Q_{13})(1 - Q_3 Q_{12})$$
$$(1 - Q_{12} Q_{13})(1 - Q_{13} Q_{23})(1 - Q_{12} Q_{23})$$

The CCF probabilities can be calculated by solving simultaneous equations in Eq. (IV–16).

By the application of this methodology of the conversion of correlated seismic failures into seismic CCFs, it becomes possible to explicitly model seismic CCFs into a MUPSA model and calculate accurate seismic CDF in a complex seismic MUPSA model with the same tools and methods that are typically used for internal, fire and flooding PSAs. As a result, this method will allow systems analysts to quantify seismic MUPSA risk as they have done with the CCF method in internal, fire and flooding PSA.

IV–3.4. Quantification uncertainties in Level 3 MUPSA

After the Fukushima Daiichi NPP accident, Level 3 PSA has emerged as an important task for assessing the risk level of the MU NPPs in a single nuclear site. Accurate calculation of the radionuclide concentrations and exposure doses to the public is required if a nuclear site has MU NPPs and a large number of people live near these NPPs. So, there has been a great need to develop a new method or procedure for fast and accurate off-site consequence calculation for MU NPP accident analysis.

Before the Fukushima Daiichi accident, Level 3 PSA had been performed only for the analysis of a SU NPP accident. Since this accident, MU Level 3 PSA has been an important task to assess the risk level of off-site consequences caused by MU NPP accidents.

Current practices for the MU Level 3 PSA assume that all the MU NPPs are located at the same position, such as a centre of mass or base NPP position, and all the radionuclides are dispersed from this single position. Consequently, the calculated off-site consequences can be distorted depending on the specific locations, MU NPP alignment and wind direction. This has been unavoidable since the current off-site risk analysis tools based on the MELCOR Accident Consequence Code System (MACCS), such as MACCS2 [IV–26] and WinMACCS [IV–27], have no capability to calculate off-site consequences from multiple NPP locations without making the above mentioned assumptions [IV–28].

In order to overcome this disadvantage of the centre of mass method, the idea of a new multiple location method was proposed [IV–29].

In the local NPP coordinate system, the x-axis represents the centreline of the plume that is a wind direction. Here, x indicates the distance from the origin (NPP) to the location along the wind direction. Radionuclides are dispersed along the plume centreline that is a wind direction. Meanwhile, y indicates the perpendicular offset distance from the plume centreline to the location where the radionuclide concentration is calculated.

MACCS2 prints out the air concentrations $C(x)$ at discrete points along the plume centreline that is a wind direction. These discrete radionuclide concentrations at ground level $(z=0)$ along the plume centreline $(y=0)$ are written to the output file. The radionuclide concentrations of the plume centreline at ground level are specified in Eq. (IV–17):

$$C(x) = \frac{Q}{\pi\sigma_y\sigma_z\bar{u}}\exp\left(-\frac{H^2}{2\sigma_z^2}\right) \tag{IV–17}$$

Here, $C(x)$ indicates the radionuclide concentration at the point x that is away from the release point (NPP), $\sigma_y(x)$ indicates the horizontal dispersion coefficient and $C(x)$ and $\sigma_y(x)$ can be calculated by the linear combination of two radionuclide concentrations and dispersion coefficients at two adjacent discrete points, which are MACCS2 (or WinMACCS) outputs on the plume centreline.

Radionuclide concentrations $C(x,y)$ at arbitrary points outside the plume centreline can be reconstructed by using a two dimensional Gaussian plume equation:

$$C(x) = \frac{Q}{\pi\sigma_y\sigma_z\bar{u}}\exp\left(\frac{y^2}{2\sigma_y}\right)\exp\left(-\frac{H^2}{2\sigma_z^2}\right)$$

C = time integrated radionuclide concentration (Bq $\times$ s/m^3)
Q = amount of radionuclide material (Bq)
H = release height (m) $\tag{IV–18}$
σ_y = horizontal dispersion coefficient (m)
σ_z = vertical dispersion coefficient (m)
$\bar{u}$ = average wind speed (m/s)

Combined radionuclide concentrations are calculated through Eq. (IV–19). Here, N is the number of NPPs. Locations (x_i,y_i) and (X,Y) denote the identical position in the local and global coordinate systems, respectively. $C(x_i,y_i)$ represents the radionuclide concentration released from the ith NPP, and $C(X,Y)$ is the radionuclide concentration released from all the NPPs in the global Cartesian coordinate system.

$$C(X,Y) = \sum_{i=1}^{N}C(x_i,\ y_i) \tag{IV–19}$$

Exposure doses can be calculated by multiplying dose conversion factors in ICRP 60 [IV–31] to radionuclide concentrations.

Both the centre of mass and multiple location methods were applied to the LOCA design basis accidents that simultaneously occur at the virtual NPPs in Fig. IV–16. Caesium-137 air concentrations that are calculated by the centre of mass and multiple location methods are depicted in Figs IV–17 to IV–20 (the information on constant weather data for virtual NPPs is presented in Table IV–9).

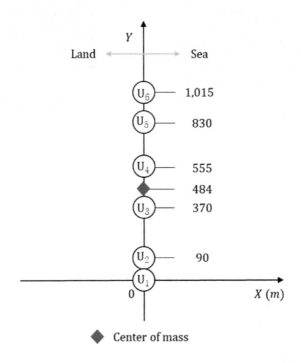

FIG. IV–16. Site layout of virtual NPP [IV–9].

TABLE IV–9. CONSTANT WEATHER DATA FOR VIRTUAL NPPs [IV–9]

Parameter	Value
Wind direction	East
Wind speed (m/s)	2.4
Precipitation (mm/h)	0
Mixing layer height (m)	1050
Atmospheric stability class	E (slightly stable conditions)

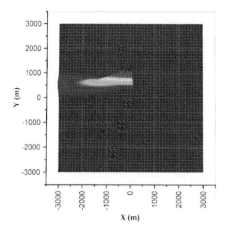

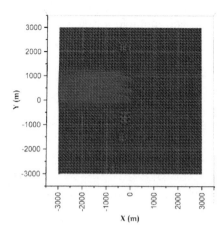

FIG. IV–17. Normalized ^{137}Cs air concentration under east wind in 2-D (centre of mass method).

FIG. IV–18. Normalized ^{137}Cs air concentration under east wind in 2-D (multiple location method).

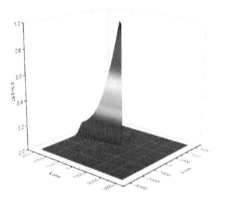

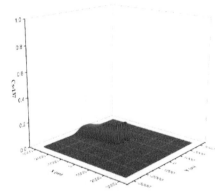

FIG. IV–19. Normalized ^{137}Cs air concentration under east wind in 3-D (centre of mass method).

FIG. IV–20. Normalized ^{137}Cs air concentration under east wind in 3-D (multiple location method).

REFERENCES TO ANNEX IV

[IV–1]   INTERNATIONAL ATOMIC ENERGY AGENCY, Nuclear Power Reactors in the World, Reference Data Series No. 2, IAEA, Vienna (2017).

[IV–2]   LIM, H.G., KIM, D.S., HAN, S.H., Development of logical structure for multi-unit probabilistic safety assessment, Nucl. Eng. Technol. 50 (2018) 1210–1216.

[IV–3]   KIM, D.S., HAN, S.H., PARK, J.H., LIM, H.G., KIM, J.H., Multi-unit Level 1 probabilistic safety assessment: approaches and their application to a six-unit nuclear power plant site, Nucl. Eng. Technol. 50 (2018) 1217–1233.

[IV–4]   CHO, J., HAN, S.H., KIM, D.S., LIM, H.G., Multi-Unit Level 2 probabilistic safety assessment: approaches and their application to a six-unit nuclear power plant site, Nucl. Eng. Technol. 50 (2018) 1234–1245.

[IV–5]   KIM, S.Y., HAN, Y.H., KIM, D.S., LIM, H., Multi-unit Level 3 probabilistic safety assessment: approaches and their application to a six-unit nuclear power plant site, Nucl. Eng. Technol. 50 (2018) 1246–1254.

[IV–6]   HAN, S.H., OH, K., LIM, H.G., YANG, J.E., AIMS-MUPSA software package for multi-unit PSA, Nucl. Eng. Technol. 50 (2018) 1255–1265.

[IV–7]   FLEMING, K.N., "On the issue of integrated risk — a PRA practitioners perspective" (Proc. Int. Topical Mtg on Probabilistic Safety Analysis (PSA 2005), San Francisco, CA, 2005), American Nuclear Society, La Grange Park, IL (2005).

[IV–8]   AMERICAN SOCIETY OF MECHANICAL ENGINEERS/AMERICAN NUCLEAR SOCIETY, Addenda to ASME/ANS RA-S–2008: Standard for Level 1/ Large Early Release Frequency Probabilistic Risk Assessment for Nuclear Power Plant Applications, ASME, New York, NY (2013).

[IV–9]   HAN, S.H., et al., "AIMS-PSA: a software for integrated PSA", in Proc. 13th Int. Conf. on Probabilistic Safety Assessment and Management (PSAM13), Seoul, Republic of Korea, 2–7 Oct. 2016.

[IV–10] NUCLEAR REGULATORY COMMISSION, Common Cause Failure Parameter Estimations, NUREG/CR-5497 (2007 Update), US NRC, Washington, DC (2007).

[IV–11] NUCLEAR REGULATORY COMMISSION, Guidelines on Modelling Common-Cause Failures in Probabilistic Safety Assessment, NUREG/CR-5485 (INEEL/EXT-97-01327), US NRC, Washington, DC (1998).

[IV–12] JUNG, W.S., et al., A fast BDD algorithm for large coherent fault trees analysis, Reliab. Eng. Syst. Saf. 83 (2004) 369–374.

[IV–13] KIM, D.-S., PARK, J.H., LIM, H.-G., A pragmatic approach to modelling common cause failures in multi-unit PSA for nuclear power plant sites with a large number of units, Reliab. Eng. Syst. Saf. 195 (2020) 106739.

[IV–14] PICKARD, LOWE AND GARRICK, Seabrook Station Probabilistic Safety Assessment — Section 13.3 Risk of Two Unit Station, PLG-0300, PLG, Irvine, CA (1983).

[IV–15] ZHANG, S., TONG, J., WU, J., "Treating common-cause failures in multi-unit PRAs" (Proc. Int. Topical Mtg on Probabilistic Safety Assessment and Analysis (PSA 2017), Pittsburgh, PA, 2017), American Nuclear Society, La Grange Park, IL (2017).

[IV–16] Kim, D.-S., Park, J.H., Lim, H.-G., "A method for considering numerous combinations of plant operational states in multi-unit probabilistic safety assessment" (Proc. 16th Int. Topical Mtg on Probabilistic Safety Assessment and Analysis (PSA 2019), Charleston, SC, 2019), American Nuclear Society, La Grange Park, IL (2019).

[IV–17] YOON, J.Y., KIM, D.-S., "A methodology for estimating the inter-unit effects of radioactive releases in multi-unit PSA", in Proc. Asian Symp. on Risk Assessment and Management (ASRAM2019), Gyeongju, Republic of Korea, 2019.

[IV–18] JEON, H., OH, K., PARK, J., "Preliminary modelling approach for multi-unit PSA in Korea [utility side]", in Proc. OECD/NEA Int. Workshop on Status of Site Level PSA (including Multi-Unit PSA) Developments, Munich, 18–20 Jul. 2018.

[IV–19] INTERNATIONAL ATOMIC ENERGY AGENCY, Safety of Nuclear Power Plants: Design, IAEA Safety Standards Series No. SSR-2/1 (Rev. 1), IAEA, Vienna (2016).

[IV–20] JEON, H., OH, K., PARK, J., "Approach to developing Level 1 multi-unit PSA models and insights from the models", in Proc. Asian Symp. on Risk Assessment and Management (ASRAM2019), Gyeongju, Republic of Korea, 2019.

[IV–21] JEON, H., OH, K., BAHNG, K., "Evaluation plan for multi-unit PSA considering operating modes in a reference site", Trans. Korean Nuclear Society Spring Mtg, Jeju, Republic of Korea, 2017.

[IV–22] JUNG, W.S., A method to improve cut set probability calculation in probabilistic safety assessment of nuclear power plants, Reliab. Eng. Syst. Saf. 134 (2015) 134–142.

[IV–23] REINERT, J.M., APOSTOLAKIS, G.E., Including model uncertainty in risk informed decision making, Ann. Nucl. Energy, 33 (2006) 354–369.

[IV–24] JUNG, W.S., ZBDD algorithm features for an efficient probabilistic safety assessment, Nucl. Eng. 239 (2009) 2085–2092.

[IV–25] JUNG, W.S., HWANG, K., PARK, S.K., "A new method to allocate combination probabilities of correlated seismic failures into CCF probabilities" (Proc. Int. Topical Mtg on Probabilistic Safety Assessment and Analysis (PSA 2019), Charleston, SC, 2019) American Nuclear Society, La Grange Park, IL (2019).

[IV–26] DEPARTMENT OF ENERGY, MACCS2 Computer Code Application Guidance for Documented Safety Analysis: Final Report, DOE-EH-4.2.1.4, 150–160, US DOE, Washington, DC (2004).

[IV–27] McFADDEN, K., BIXLER, N.E., CLEARLY, V.D., EUBANKS, L., HAAKER, R., User's Guide and Reference Manual WinMACCS Version 3, US NRC, Washington, DC (2007).

[IV–28] NUCLEAR REGULATORY COMMISSION, RASCAL 4.3: Description of Models and Methods, NUREG-1940, Washington, DC (2015).

[IV–29] LEE, H.R., LEE, G.M., JUNG, W.S., A method to calculate off-site consequences for multi-unit nuclear power plant accident, J. Korean Soc. Saf. 336 (2018) 144–156.

[IV–30] INTERNATIONAL COMMISSION ON RADIOLOGICAL PROTECTION, The 1990 Recommendations of the International Commission on Radiological Protection, Publication 60, Pergamon Press, Oxford and New York (1991).

Annex V

UNITED KINGDOM — ASSESSING AND REGULATING
SAFETY ON MU FACILITY SITES

The United Kingdom (UK) has a wide range of nuclear sites and facilities; however, the content of this annex will focus on those aspects of assessment and regulation of safety on multifacility sites containing a civil NPP relevant to the topic of PSA.

V–1. BACKGROUND

Figure V–1 shows the location of civil NPP sites in the UK. From the earliest days of civil nuclear power, the UK's gas cooled reactors were built in pairs, and this continued throughout the development of first the Magnox and then the advanced gas cooled reactor NPPs. Sizewell B, a PWR and the most recent civil NPP in the UK, is the only single NPP site in the UK. However, Sizewell B is sited next to Sizewell A, two Magnox reactors that are now in decommissioning, and the Sizewell C site where EDF Energy plans to construct two EPR type NPPs.

All planned new build projects involve multiple units on the same site, and all sites identified by the UK for development of new large NPPs [V–1] are adjacent to existing sites. Small modular reactors and advanced modular reactors are also being considered for future development and deployment in the UK [V–2] and are expected to rely upon the deployment of multiple reactor modules on the same site to make the economic case for development.

There are many other nuclear licensed sites containing facilities other than NPPs in the UK, with some having many varied types of facilities. For example, Sellafield contains multiple facilities on the same site, including waste processing and storage facilities, reactors undergoing decommissioning and legacy waste facilities. Sellafield is a complex site where safety and regulation need to be considered from a multifacility point of view.

Regulation of nuclear facilities in the UK is the responsibility of the ONR and includes the regulation of safety, security, transport and safeguards; the environmental agencies[1] are responsible for regulating environmental aspects,

---

[1] Each of the three countries comprising Great Britain has its own environmental agency: the Environment Agency in England, National Resources Wales in Wales and the Scottish Environment Protection Agency in Scotland.

such as permitting (authorizing) normal atmospheric and aquatic discharges. Safety is the responsibility of the licensee operating the facility, which needs to ensure that risks are reduced to a level as low as reasonably practicable (ALARP) in order to comply with the law.

Potential new reactor designs are assessed by the UK regulators under the generic design assessment (GDA) process [V–3], which was developed jointly by the regulators for assessing new reactor designs. GDA is multistep process that to date has taken around four to six years. It is initiated at the request of the Government following an application by a requesting party (e.g. the reactor vendor). The process involves assessment by the regulators of submissions covering safety, security and the environment for construction and operations of the proposed reactor design at a generic UK site. The scope of the assessment is agreed between the requesting party and the regulators. Successful completion of a GDA, however, is not sufficient to begin construction of the NPP. Construction can start only when a nuclear site licence has been granted, consent to construct has been given by ONR and any other relevant permits, permissions and consents have been granted. The regulatory decision to grant a nuclear site licence and consent to construct will follow assessment of a safety case submitted by the licensee [V–4].

V–2. SAFETY REGULATION

The UK's Health and Safety at Work etc. Act 1974[2] is overarching and applies to most work activities, including operation of NPPs. The requirements of this Act are based on ensuring that risks to workers and the public are reduced so far as is reasonably practicable (often substituted by ALARP).

The UK's regulatory system for nuclear safety is goal-setting and non-prescriptive and so does not generally include prescribed requirements in terms of limits or methodologies. However, ONR does set out its expectations as guidance for its staff when making assessment decisions in the form of SAPs [V–5] supported by Technical Assessment Guides for individual subject areas such as PSA [V–6]. This guidance includes the numerical targets within the SAPs for radiological risks to workers, individual members of the public and society from faults.

The UK's approach to nuclear safety regulation is summarized in the ONR's Risk Informed Regulatory Decision Making [V–7] and is not reproduced in detail in this annex.

---

[2] https://www.legislation.gov.uk/ukpga/1974/37/contents

UK NUCLEAR SITES...

The UK has a wide range of **nuclear expertise**, from fuel fabrication, operating nuclear power stations and decommissioning. The sites are located across the country, and employ **over 63,000 people.**

Dounreay
Thurso, Caithness

Key

- 🕮 Advanced Gas-cooled Reactor (AGR)
- 🅂 Fuel plant
- 🕮 Decommissioning sites
- ⊙ Fusion research
- 🏯 Pressurised Water Reactor (PWR)
- 🕮 Proposed new build sites

Hunterston A Power Station
Hunterston B Power Station
West Kilbride, Ayrshire

Torness Power Station
Dunbar, East Lothian

Chapelcross Power Station
Annan, Dumfriesshire

Moorside Power Station
West Cumbria

Sellafield (Decommissioning)
Seascale, Cumbria

Hartlepool Power Station
Hartlepool, Cleveland

Low Level Waste Repository
Near Drigg, Cumbria

Heysham 1&2 Power Stations
Morecambe, Lancashire

Springfields (Fuel Manufacturing)
Springfields
Near Preston, Lancashire

Wylfa Newydd Power Station
Wylfa Power Station
Cemaes, Anglesey

Capenhurst (Fuel Manufacturing)
Capenhurst (Decommissioning)
Capenhurst, Cheshire

Trawsfynydd Power Station
Blaenau Ffestiniog, Gwynedd

Culham Centre
for Fusion Energy
Oxfordshire

Berkeley Power Station
Berkeley, Gloucestershire
Oldbury Power Station
Oldbury Power Station
Oldbury, Gloucestershire

Sizewell A Power Station
Sizewell B Power Station
Sizewell C Power Station
Near Leiston, Suffolk

Hinkley Point A Power Station
Hinkley Point B Power Station
Hinkley Point C Power Station
Bridgwater, Somerset

Bradwell Power Station
Southminster, Essex

Dungeness A Power Station
Dungeness B Power Station
Romney Marsh, Kent

Winfrith
Winfrith Newburgh, Dorset

Harwell
Harwell, Oxfordshire

FIG. V–1: Figure reproduced from Nuclear Industry Association booklet 'Nuclear Energy Facts' (https://www.niauk.org/nuclear-energy-facts/). Bradwell is also a proposed new build site.

V–3. PSA — REGULATORY EXPECTATIONS

PSA was embraced quite early in the UK and was included in regulatory expectations from the 1970s. Licensees and operators are responsible for developing and maintaining PSAs, with models and results submitted to the regulator.

ONR's SAPs set out the expectation that 'suitable and sufficient' PSA should be performed as an integral part of the design development and fault analysis, with the scope and depth of the PSA being proportionate to the magnitude of the radiological hazard and risks, the novelty of the design, the complexity of the facility and the nature of the decision that the PSA is supporting. For a complex facility, such as a NPP, a comprehensive PSA is expected.

The main SAPs concerning PSA are the following:

(a) Fault analysis should be carried out, comprising suitable and sufficient design basis analysis, PSA and severe accident analysis, to demonstrate that risks are ALARP;

(b)   Suitable and sufficient PSA should be performed as part of the fault analysis and design development and analysis;

(c)   PSA should reflect the current design and operation of the facility or site;

(d)   PSA should cover all significant sources of radioactivity, all permitted operating states and all relevant initiating faults;

(e)   The PSA model should provide an adequate representation of the facility and/or site.

While MUPSA is not explicitly required, it may be considered to be part of a 'suitable and sufficient' PSA, and the scope of the PSA developed for each site or facility needs to be justified on a case by case basis. SAPs also expect that

"[T]he identification of initiating faults should consider the potential for combinations of hazards. At multi-facility sites, the analysis should also consider the potential for specific initiating faults giving rise to simultaneous impacts on several facilities or for faults in one facility to impact another facility."

The SAPs recognize that facilities built to earlier standards may not always be able to meet modern standards and that the issue of whether suitable and sufficient measures are available to satisfy risks being reduced to ALARP will need to be judged on a case by case basis.

For the new reactor designs undergoing GDA, ONR has made the following statement regarding PSA in its guidance to requesting parties for GDA [V–3]:

"ONR expects that the GDA submissions will include a full scope Level 1 and Level 2 PSA. The PSA should be used to help show that the design satisfies the ALARP requirement. A Level 3 PSA relevant to the generic site will also be expected."

Detailed guidance on PSA is provided in the PSA Technical Assessment Guide [V–6] and Technical Guidance for GDA [V–8], which are intended as guidance for ONR inspectors assessing safety submissions. The PSA Technical Assessment Guide has recently been updated to include a section on MUPSA, following international engagement with MUPSA projects at the IAEA and the OECD/NEA Working Group on Risk Assessment, with the following high level guidance to inspectors:

"[D]utyholders are expected to consider the impact of multiple reactor units and other facilities within the PSA. For existing sites with limited dependencies between individual facilities this may be limited to a single

unit PSA for a representative reactor unit or facility, supported by a justification for this approach providing a suitable and sufficient PSA to support the expected PSA applications. For a new build site, or a site with significant or complex interactions between facilities, more detailed analysis may be required."

V–4. PSA — UK EXPERIENCE

Use and development of PSA at operating UK NPPs varies. All NPP PSAs developed for the UK to date have been for a SU, with assumptions and bounding calculations made to represent additional units as required.

One of the first full scope PSAs — all operating states and Levels 1, 2 and 3 — was performed for Sizewell B licensing in the early 1990s, and the Level 1 PSA continues to be maintained as a living PSA. PSAs for the advanced gas cooled reactors are limited to Level 1 PSA for at-power operation, with additional bespoke studies performed for internal hazards, shutdown states and Level 2 PSA.

Sizewell B and the newer advanced gas cooled reactors use the PSA as the basis for a risk monitor to risk inform operational decisions and maintenance and outage planning.

Recent PSAs submitted as part of GDAs or applications for a nuclear site licence or consent to construct have been full scope PSAs, considering aspects such as internal and external hazards, the SFP and reactor/SFP combinations and to Level 2 and 3.

As discussed above, the scope of the GDA is agreed between the requesting party and the regulators and may include multiple units; however, all GDAs to date have been for SUs. Successful completion of a GDA does not allow for construction of the NPP to begin. Construction can start only when a nuclear site licence has been granted and consent to construct is given by ONR. The regulatory decision on granting a nuclear site licence and consent to construct will follow assessment of a safety case submitted by the licensee, which needs to consider any site specific aspects, such as the number of units planned to be constructed on the site, any neighbouring facilities and any proposed future construction [V–7].

An example of this is the Hinkley Point C project, where the PSA has been further developed after successfully completing GDA to seek consent for nuclear island construction (granted in 2018) and later milestones. From the MUPSA perspective, development of the Hinkley Point C PSA since GDA has included:

(a) Turbine disintegration: the impact due to turbine disintegration has been assessed, taking into account two turbines associated with the proposed twin

EPR reactors on the Hinkley Point C site and also the two turbines on the existing Hinkley Point B site;

(b)   Radiological release from Hinkley Point B has been captured as an event in the PSA;

(c)   A simplified assessment of the level of risk from the twin-unit site has been performed (multiplying the single PSA risk results by a factor of two);

(d)   A twin reactor PSA strategy, including a plan by EDF R&D to propose a methodology and an item captured in the PSA forward work plan.

V–5.   SAFETY GOALS

The SAPs [V–5] include numerical targets that apply to all types of facilities.[3] For the safety analysis of an NPP, a PSA would be expected to produce results for comparison with Targets 7–9, which concern radiological risks to the public from faults and provide an input into Targets 5 and 6, which concern radiological risk to workers from faults.

— Target 5 — Individual risk of death from accidents at the **site** for any person on the site;
— Target 6 — Frequency dose targets for any single accident in the **facility** for any person on the site;
— Target 7 — Individual risk to people off the site from accidents at the **site**;
— Target 8 — Frequency dose targets for accidents on an individual **facility** for any person off the site;
— Target 9 — Total risk of 100 or more fatalities from accidents at the **site** (either immediate or eventual and on- and off-site), with consequences truncated at 100 years and limited to the effects on the UK population.

Each target has two levels: basic safety levels and basic safety objectives. The basic safety levels are higher than the basic safety objectives and indicate doses/risks that new facilities have to meet, in addition to being shown to be ALARP, and they provide benchmarks for existing facilities. The basic safety objective doses/risks have been set at a level at which ONR considers it not to be a good use of its resources or taxpayers' money, nor consistent with a targeted and proportionate regulatory approach, to pursue further improvements in safety. Facility operators and owners have an overriding duty to consider whether they have reduced risks to ALARP on a case by case basis, irrespective of whether the basic safety objectives are met.

---

[3] Annex 2 of the SAPs describes the basis and derivation of the numerical targets.

SAPs Numerical Targets 5, 7 and 9 relate to site wide risk, and therefore assessment against these targets has to include consideration of all facilities on the site. The SAPs contain a definition for societal risk (Target 9), which refers to "an activity from which risk is assessed as a whole and is under the control of one company in one location, or within a site boundary".

SAPs Numerical Targets 6 and 8 relate to the risk arising from a facility. The SAPs define a facility as "A part of a nuclear site identified as being a separate unit for the purposes of nuclear or radiological risk". From a PSA point of view, dutyholders are expected to define the boundaries of each facility for the purposes of assessment against Targets 6 and 8. For example, at a traditional NPP site each unit may be treated as a separate facility. However, where individual units are coupled or connected, or where multiple reactors are contained in close proximity within the same building or the same body of water, they may be considered to be the same facility for the purposes of assessment against Targets 6 and 8.

For the safety analysis of a NPP, a PSA would be expected to produce results for comparison with Targets 7–9. Target 8 generally requires a hybrid probabilistic-deterministic assessment: a Level 1 and 2 PSA to evaluate the off-site release frequencies and then a deterministic assessment to calculate the consequences. Targets 7 and 9 may require a full Level 1–3 PSA with a probabilistic assessment of the off-site consequences, taking into account the effects of different meteorological conditions and their frequencies at Level 3.

Target 9 is site based, and therefore requires all facilities on a site to be considered. Experience to date has shown that the consequences predicted for a SU release are dominated by long term effects in the general population resulting from consumption of food below the Community Food Intervention Levels set by the EU [V–9] (i.e. food considered safe to eat) or from other exposure at low doses of large populations at some distance from the plant.

Given that some targets are site based, consideration of MU effects may have an impact on the justification of total site risk, with consideration of Target 9 potentially being complex given the dominant effect of a large population being exposed to relatively low doses.

The presence of multiple units on the same site may only have a limited impact on the assessment of consequences against Target 9 because of saturation effects (i.e. a two-unit release has comparable consequences to a SU release). On the other hand, the impact of multiple 'small' releases (not individually meeting the Target 9 criterion of 100 fatalities) from the same site may have a non-linear effect on the consequences predicted and therefore on the site risk assessment against Target 9. This non-linear effect may also have an impact on the assessment of site risk against Target 7 and is a topic of ongoing research within ONR, as described in the next section.

V–6. RESEARCH AND ONGOING WORK

Since the Fukushima Daiichi accident, a number of research activities have been undertaken by the UK nuclear industry and ONR on the topic of MUPSA. The research undertaken by ONR is summarized in this section.

ONR has engaged with international research activities on MUPSA, including work at IAEA to develop a MUPSA methodology (presented in this safety report) and the OECD/NEA Working Group on Risk Assessment to understand the international state of practice and share international experience [V–10]. These activities informed the 2019 update of the ONR Technical Assessment Guide on PSA [V–8], which includes a new section on MUPSA PSA expectations as described above.

ONR also launched specific research on Level 3 MUPSA, to increase understanding of MU impacts on the off-site dose consequence analysis. A summary of this research has been published [V–11] and will be taken into account in the future development of MUPSA regulatory expectations. The research used source term data provided by the industry and ONR to explore the impact of different combinations of simultaneous and time offset releases at two sites in the UK.

The research concluded that an additive combination (for individual risk) or linear combination (for societal risk) of SU results provided a good approximation for simultaneous MU releases. However, more complex effects were seen for time offset releases and some areas of potential non-conservatism for deterministic consequences. A number of recommendations were made, including areas where sensitivity studies on a site specific basis may provide further information. It is important to recognize that the research was simplified and generic in some aspects and, therefore, cannot be directly compared to any NPP or site. Consequently, additional site specific analyses may be required on a case by case basis.

REFERENCES TO ANNEX V

[V–1] DEPARTMENT OF ENERGY AND CLIMATE CHANGE, National Policy Statement for Nuclear Power Generation (EN-6), The Stationery Office, London (2011),
https://www.gov.uk/government/publications/national-policy-statements-for-energy-infrastructure

[V–2] DEPARTMENT FOR BUSINESS, ENERGY & INDUSTRIAL STRATEGY, Industrial Strategy: Nuclear Sector Deal, BEIS, London (2018),
https://www.gov.uk/government/publications/nuclear-sector-deal

[V–3] OFFICE FOR NUCLEAR REGULATION, New Nuclear Reactors: Generic Design Assessment Guidance to Requesting Parties ONR-GDA-GD-001 Revision 3, ONR, Bootle, UK (2016),
http://www.onr.org.uk/new-reactors/guidance-assessment.htm

[V–4] OFFICE FOR NUCLEAR REGULATION, Licensing Nuclear Installations, 4th edn, ONR, Bootle, UK (2015),
http://www.onr.org.uk/licensing-nuclear-installations.pdf

[V–5] OFFICE FOR NUCLEAR REGULATION, Safety Assessment Principles for Nuclear Facilities, ONR, Bootle, UK (2014),
http://www.onr.org.uk/saps/

[V–6] OFFICE FOR NUCLEAR REGULATION, Nuclear Safety Technical Assessment Guide, Probabilistic Safety Analysis, NS-TAST-GD-030 Revision 6, ONR, Bootle, UK (2019).

[V–7] OFFICE FOR NUCLEAR REGULATION, Risk Informed Regulatory Decision Making, ONR, Bootle, UK (2017).

[V–8] OFFICE FOR NUCLEAR REGULATION, New Nuclear Power Plants: Generic Design Assessment Technical Guidance, ONR-GDA-GD-007, ONR, Bootle, UK (2019),
http://www.onr.org.uk/new-reactors/reports/onr-gda-007.pdf

[V–9] EUROPEAN ATOMIC ENERGY COMMUNITY, European Union Directive 87/3954/EURATOM (as amended by 89/2218/EURATOM and supplemented by Commission Regulation 89/944/EURATOM), European Commission, Brussels (1987).

[V–10] ORGANISATION FOR ECONOMIC CO-OPERATION AND DEVELOPMENT/NUCLEAR ENERGY AGENCY/COMMITTEE ON THE SAFETY OF NUCLEAR INSTALLATIONS, Status of Site Level (Including Multi-Unit) PSA Developments, NEA/CSNI/R(2019)16, OECD/NEA/CSNI, Paris (2019).

[V–11] OFFICE FOR NUCLEAR REGULATION, Research Report on Multi-Unit Level 3 PSA — Executive Summary, ONR-1908/ONR-RRR-084, Jacobsen Analytics, ONR, Bootle, UK (2019),
http://www.onr.org.uk/documents/2019/onr-rrr-084.pdf

Annex VI

UNITED STATES OF AMERICA — APPROACH TO MUPSA

This annex includes white papers on the following topics:

— VI–1. NRC: MU risk considerations for operating reactors and advanced LWRs;
— VI–2. EPRI: First phase of a project on risk-informing MU risk aspects using PSA models;
— VI–3. Scoping approach for MUPSA.

VI–1. NRC: MU RISK CONSIDERATIONS FOR OPERATING REACTORS AND ADVANCED LWRS

In the United States of America (USA), there are 32 sites with two operating LWRs and three sites with three operating LWRs. When Vogtle Unit 3 and Unit 4 (two AP1000 units) become operational (planned for 2023), there will be one site with four reactor units. Another 22 sites operate with only one reactor unit. With few exceptions, PSAs of NPPs in the USA have focused on developing PSA models for SUs. All PSA models, however, have modelled some key aspects of MU impacts (e.g. site wide initiators such as LOOP and incorporated shared systems to the PSA models after an appropriate treatment of plant procedures). NRC regulations, however, have always recognized the MU accident risk considerations in design and the policy considerations in a risk informed performance manner.

Some examples of regulatory treatment of MU risk considerations in design and siting are as follows:

10 CFR Part 50, Appendix A, General Design Criterion 5 [VI–1] states:

"Structures, systems, and components important to safety shall not be shared among nuclear power units unless it can be shown that such sharing will not significantly impact their ability to perform their safety functions, including, in the event of an accident in one unit, an orderly shutdown and cooldown of the remaining units."

10 CFR 100.11(b) [VI–2], which provides requirements for determining the exclusion area, the low population zone and the population centre distance for MU sites, states:

"Subsection (b)(1): If the reactors are independent to the extent that an accident in one reactor would not initiate an accident in another, the size of the exclusion area, low population zone and population centre distance shall be fulfilled with respect to each reactor individually.

Subsection (b)(2): If the reactors are interconnected to the extent that an accident in one reactor could affect the safety of operation of any other, the size of the exclusion area, low population zone and population centre distance shall be based on the assumption that all interconnected reactors emit their postulated fission product releases simultaneously."

Some examples of NRC Commission views with respect to MU risk considerations for the operating reactors are as follows.

In response to the accident at Three Mile Island, the NRC issued an action plan (NUREG-0660) [VI–3]. Item II.B.8 involved a two phase rulemaking proceeding on degraded core accidents. In the second phase (termed the 'long-term rulemaking'), the NRC identified the need to consider the effects of an accident in a reactor plant on an adjacent plant in a multiple reactor site. This issue was subsequently dropped at the Commission's direction, as discussed in the Staff Requirements Memorandum to SECY-82-1B [VI–4]:

"There are other issues listed in Item II.B.8 of NUREG-0660 that the Commission believes have minimal value for improved safety and, therefore, need not be considered further: namely, effects of severe accidents at multi-unit sites and post-accident recovery plans."

Moreover, the Commission's safety goals, which define acceptable risk, are applied on a per reactor basis. NUREG-0880 [VI–5] summarizes comments made by the public as the safety goals were being formulated in the early 1980s:

"Some commenters objected to the originally proposed individual and societal numerical guidelines because they were to be applied on a per-site basis. This would have resulted in tighter requirements being imposed on plants at multi-unit sites than at single-unit sites. The Commission decided not to impose a regulatory bias against multi-unit sites. Therefore, the quantitative design objectives were changed from risks per site to risk per plant."

Because of the Commission's decision to apply the safety goals on a per reactor basis, as well as advances in the state of the art (e.g. lower than previously estimated source terms, slower progression of core damage accidents due to accident mitigation strategies) and plant safety enhancements due to post-Fukushima modifications, there has been no regulatory impetus to estimate the total risk of a MU site for US operating reactors.

VI–1.1. Enhancement to safety due to post-Fukushima orders

Operating experiences [VI–6], as well as numerous other site wide LOOP events, provide proof of MU reactor accidents and the possible need to consider those risks in regulatory policy and decision making. The sequence of events during the Fukushima Daiichi accident has been reviewed extensively by the nuclear industry and the NRC [VI–7, VI–8]. That accident clearly demonstrated that MU accidents are credible, and their risk can be an important contributor to the overall risk associated with the site.

Following the Fukushima Daiichi accident, the NRC established a Near-Term Task Force in response to a Commission direction. The Near-Term Task Force developed a comprehensive set of recommendations and issued it to the Commission on 12 August 2011 [VI–7]. The subject report included several recommendations by a panel of NRC experts. Recommendation 2.3 stated that licensees have to be ordered to perform seismic and flooding walkdowns to identify plant specific vulnerabilities against the current design basis. Recommendation 2.1 stated that licensees have to be ordered to evaluate the seismic and flooding hazards at their sites against the NRC requirements and guidance imposed on new reactors. Consequently, the Commission issued directions outlining a set of near term and long term actions that have to be implemented by the staff via staff requirement memorandum SECY-11-0124 [VI–9]. The NRC eventually sent letters under 10 CFR 50.54(f) to its licensees, requesting that they address recommendations 2.1 and 2.3 for seismic and flooding hazards [VI–10]. The NRC also issued an order EA-12-049 [VI–11] to modifying licences with regard to requirements for mitigation strategies for beyond design basis external events. This order requires a three phase approach for mitigating beyond design basis external events. The initial phase requires the use of installed equipment and resources to maintain or restore core cooling, containment and SFP cooling capabilities. The transition phase requires provision of sufficient, portable on-site equipment and consumables to maintain or restore these functions until they can be accomplished with resources brought from off the site. The final phase requires obtaining sufficient off-site resources to sustain those functions indefinitely.

The order also stated that these strategies need to be capable of mitigating a simultaneous loss of all AC power and loss of normal access to the UHS and need to have adequate capacity to address challenges to core cooling, containment and SFP cooling capabilities at all units on a site.

In response to this order, the US nuclear power industry made various changes to its plants (e.g. procurement and sometimes installation of additional components, development of procedures) to address the Commission's orders. Each plant installed new emergency response equipment that is stored on-site and protected from natural hazards. NRC inspectors have verified that the strategies are in place at all the US NPPs. The nuclear power industry also established two national centres to store additional equipment that can be used to supplement the on-site equipment if needed.

Even though the orders were issued to induvial reactor units, the modifications that were required to comply with the orders were focused primarily on MU accidents such as loss of all AC and heat sink. Consequently, the MU risk component for all US sites is lower compared to its level prior to the Fukushima event.

VI–1.2. Margins between quantitative safety goals and risk metrics

The NRC established its safety goal policy in 1986 [VI–12]. In that policy, the NRC established two qualitative safety goals and supporting QHOs. The qualitative safety goals were that the operation of NPPs would pose no significant additional risk to an individual, and that the risks to society would be comparable to or less than those associated with other forms of generating electricity. The associated QHOs were as follows:

> *"The risk to an average individual in the vicinity of a nuclear power plant of prompt fatalities that might result from reactor accidents should not exceed one-tenth of one percent (0.1%) of the sum of prompt fatality risks resulting from other accidents to which members of the US population are generally exposed.*
>
> *The risk to the population in the area near a nuclear power plant of cancer fatalities that might result from nuclear power plant operation should not exceed one-tenth of one percent (0.1%) of the sum of cancer fatality risks resulting from all other causes."*

For regulatory decision making, NRC uses a surrogate to the subsidiary objectives, also referred to as a 'risk metric'. For plants licensed under 10 CFR 50 and 10 CFR 52 (e.g. AP1000), the risk metric for CDF is 1×10^{-4}/year. For plants licensed under 10 CFR 50, the risk metric for LERF is 1×10^{-5}/year. For plants licensed under 10 CFR 52, the risk metric for LERF is 1×10^{-6}/year. Appendix D of NUREG-1860 [VI–13] used risk analyses performed by NRC in the late 1980s and documented in NUREG-1150 [VI–14] to correlate the QHOs with risk metrics.

In 2018, the EPRI performed a technical evaluation of the margins between QHOs and the risk metrics using various NRC studies that had demonstrated that (a) the amount of radioactivity release during a core damage event is substantially less than previously assumed, and (b) the rate of core melt progression is slower than previously assumed (hence providing additional time to evacuate). Using these insights, EPRI stated that there are significant margins (there is an order of 10 000 between LERF and early fatality risk and an order of 100 between CDF and latent fatality risk) [VI–15].

NRC staff, when informed of the EPRI report (i.e. NRC has not performed a formal review of the report), informed EPRI and the industry that (a) QHOs are a subset of the safety goal policy (a narrow focus on the numerical margins ignores the intention of the entirety of NRC's safety goal policy), (b) the entirety of NRC responsibilities, rules and policies needs to be considered when translating numerical margins to regulatory margins, and (c) staff need to carefully consider the applicability of past studies for the purpose of estimating numerical margins when translating them to risk informed decision making.

The NRC staff did not, however, refute the fact that there is more margin than was assumed when appendix D of NUREG-1860 was prepared in the 1980s, since the follow up work, such as the state of the art reactor consequence analyses project [VI–16], has shown that core melt progression will be slower than estimated and that source terms will be smaller than estimated.

VI–1.3. Summary of the planned screening approach that staff will use to consider MU risk consideration

While recognizing that MU risks do not pose safety concerns, the NRC is of the view that MU risk considerations need to be included, qualitatively or quantitatively, in risk informed decision making. Therefore, the NRC staff is in the process of using (a) analytical work relating to MU risks, (b) insights gleaned from MUPSAs and SFP risk assessments that have already been performed, and (c) the Level 3 PSA project that is under way at the NRC to develop a screening methodology to enable the NRC to glean risk insights at a site specific level. The paragraphs below provide a synopsis reflective of the current status of that effort.

In 2012, a paper entitled 'An event classification schema for evaluating site risk in a multi-unit nuclear power plant probabilistic risk assessment' [VI–17] provided a number of insights into developing an informed analysis to include risk considerations for sites with two to three units. That paper acknowledged that PSAs at MU NPPs need to consider risk from each unit separately, as well as considering dependencies and interactions between the units informally and on an ad hoc basis. The paper highlighted the fact that the accident at the Fukushima Daiichi nuclear power station underlined the importance and possibility of MU accidents. The paper then provided six main factors (dependence classifications) that have to be effectively accounted for in assessing MU considerations. These are:

— IEs;
— Shared connections;
— Identical components;
— Proximity dependencies;
— Human dependencies;
— Organizational dependencies.

In 2014, in response to some NRC staff questions about the magnitude of MU risk, NRC undertook an initiative to develop a methodology to arrive at bounding estimates of MU risks for reactors located on a given site. That bounding methodology is documented in a paper entitled 'Scoping estimates of multiunit accident risk' [VI–18].

From Ref. [VI–18], when applied to three-unit site, the total risk results in a scoping estimate of:

$$R_{site,3} \leq 3R_{cci,1} + 9R_{sui,1}$$

Here,

$R_{site,3}$ is the total risk from the three-unit site.

$R_{cci,1}$ is the risk from a SU due to a common initiator such as an earthquake.

$R_{sui,1}$ is the risk from a SU due to SUIEs such as a LOCA.

As discussed before, the plant modifications implemented by US plants to meet additional requirements imposed on them because of the Fukushima Daiichi accident prompted reductions in the risk associated with common initiators $\{R_{cci,1}\}$. For plants whose risk profile was dominated by loss of power events, these reductions in risk were substantial.

Based on the above, it is not unreasonable to expect that the risk at a site with three units is unlikely to be an order of magnitude larger than the risk associated with one unit. In fact, unless there are vulnerabilities that enable a release from one unit to cause difficulties for the operators of other units to safe shutdown the reactor, or cause relatively high CCF across units, substantial increases in risk are not likely. Because of this, and in consideration of the margins that exist between the QHOs, the NRC safety goals and the risk metrics used to support risk informed decision making, there is no impetus to require US licensees with two or three reactor units to develop detailed MUPSA models. The NRC will continue to evaluate MU risk and plans to develop a method that adequately captures MU risk consideration for use in risk informed decision making.

VI–1.4. MU risk consideration for non-LWRs or advanced reactors with more than one module

More recently, the NRC issued Revision 3 to Chapter 19 of NUREG-0800 [VI–19] to include considerations for multimodule risk. This update states:

"For small, modular integral pressurized water reactor designs, the staff reviews the results and description of the applicant's risk assessment for a single reactor module; and, if the applicant is seeking approval of an application for a plant containing multiple modules, the staff reviews the applicant's assessment of risk from accidents that could affect multiple modules to ensure appropriate treatment of important insights related to multi-module design and operation.

The staff will verify that the applicant has:

i. Used a systematic process to identify accident sequences, including significant human errors, that could lead to multiple module core damages or large releases and described them in the application

ii Selected alternative features, operational strategies, and design options to prevent these sequences from occurring and demonstrated that these accident sequences are not significant contributors to risk. Operational strategies should also provide reasonable assurance that there is sufficient ability to mitigate multiple core damages accidents."

To support the future licensing of non-LWR plants, the US nuclear power industry is developing technology-inclusive, risk informed and performance based guidance for the selection of licensing basis events, the classification of SSC and the evaluation of defence in depth [VI–20]. This guidance states:

"If applicable, the PRA should include event sequences involving two or more reactor modules as well as two or more sources of radioactive material.

This enables the identification and evaluation of risk management strategies for reactor modules and sources within the scope of a single application to ensure that sequences involving multiple reactor modules and sources are not risk significant."

In addition, ASME and ANS have developed a trial use standard for the performance of PSAs for advanced non-LWR NPPs [VI–21].

VI–1.5. Research activities

In the research arena, the NRC Office of Nuclear Regulatory Research is in the intermediate stages of an effort to create an integrated Level 3 PSA that includes the effects of multiple units, as well as the risk from all radiation sources on-site, such as the SFP. The ongoing integrated site Level 3 PSA study is being performed for research purposes. Although some potential future uses within the NRC's regulatory framework have been identified, this Level 3 PSA is not intended to support a specific risk informed application. Instead, the fundamental objectives of this study are to [VI–22, VI–23]:

(a) Develop a contemporary Level 3 PSA generally based on current state of practice methods, models, data and analytical tools that:
  (i) Reflects technical advances since the last NRC-sponsored Level 3 PSAs were performed as part of the NUREG-1150 study;
  (ii) Addresses risk contributors not previously considered, including concurrent accidents involving multiple co-located radiological sources (i.e. MU accidents).
(b) Extract new risk insights to:
  (i) Enhance regulatory decision making;
  (ii) Help focus limited resources on issues most directly related to NRC's mission to protect public health and safety.
(c) Enhance NRC staff's internal PSA capability and expertise;
(d) Improve PSA documentation to make information more accessible, retrievable and understandable;
(e) Obtain insight into the technical feasibility and cost of developing new Level 3 PSAs.

Although this Level 3 PSA study is generally being performed in a manner consistent with current standards and state of practice using existing PSA technology, there are some technical elements that necessitate methodological development because of a lack of sufficient experience to define a current state of practice. One such technical element is the Integrated Site PSA (i.e. MU or multisource PSA) technical element that is being developed in support of, and then applied to, the Level 3 PSA study. More detailed information on the Level 3 PSA study approach for Integrated Site PSA will be available at the completion of the study.

VI–2. EPRI FIRST PHASE OF A PROJECT ON RISK-INFORMING MU RISK ASPECTS USING PSA MODELS

VI–2.1. Introduction

This section summarizes the initial findings of a phased project to develop a framework for assessing MU risk at NPPs by EPRI. At the outset of this project, EPRI recognized that, while approaches and methodologies have been previously proposed to address MU risk, their feasibility and implementation using current tools within a risk informed decision making (RIDM) process are still evolving (despite some applications discussed in other annexes of this publication). Hence, EPRI engaged in an effort to apply and extend existing approaches, evaluate their feasibility and identify remaining challenges and technical gaps from a practical perspective. A report with the complete details has been recently published [VI–24].

With regard to this IAEA safety report, although there was no explicit intent to pilot the methodology, several aspects of the EPRI effort were aligned in specific areas. For example, the effort recognized that existing PRA models may need to be refined with regard to MU risk aspects, i.e. dependencies and cross-ties that may be implicitly or explicitly modelled in SUPSA models may need to be modified for use in MU risk assessment. The use of operating experience data to potentially screen specific combinations of at-power and LPSD modes based on frequency of occurrence of such combinations was also explored. In addition, an iterative approach to modelling CCF and HRA (including simplified/conservative assumptions) was pursued in the EPRI effort, with the recognition that these aspects can get overly complicated as well as limited by available data. Finally, as indicated in the IAEA case study, specific hazards such as seismic are also highlighted as potentially becoming the main contributors to the MU risk profile.

VI–2.2. Project overview

This EPRI project includes leveraging the existing PSA models from three different dual-unit pilot sites where MU risk modelling has not been performed previously. Dual-unit sites with Generation II designs for PWRs are the focus of the initial phases of this effort, to explore in more detail the PSA-intensive aspects of MU risk (i.e. before extending into more challenging and resource intensive variations).

Two general approaches were defined to characterize the boundaries of a graded framework to address MU risk. The first is the detailed MUPSA approach. This includes the development of a comprehensive PSA-focused approach to deriving risk insights for aspects specific to multiple units on a plant site, and is a more explicitly quantitative modelling approach to MU risk because it combines and extends logic models in SUPSAs to obtain MU risk numerical insights. The second approach is the simplified MU risk assessment approach. In this case, approximate methods and qualitative insights based on using existing SUPSAs to estimate or bound MU risk will be the focus. The development of the simplified approach requires an a priori understanding of the important MU risk aspects at the specific site and an extrapolation of the risk insights via bounding and/or screening approaches. The basic premise of this effort is that a simplified approach may be an acceptable part of RIDM approaches as long as limitations are understood and are appropriate for the intended application. In some cases, a detailed approach to a specific portion of the MU risk assessment (e.g. Level 1 PSA results for select internal event initiators) may be used in combination with simplified approaches to other aspects of the MU risk model.

This is intended to recognize that (a) not all States may require a full scope PSA to address MU aspects, (b) regulatory regimes may vary in terms of expectations for MU requirements, (c) there is great variation in terms of aspects that may impact on MU risk (for example, MU risk may be more relevant for sites with a large number of units co-located, which are a minority in current large commercial reactor applications) and (d) the resources to develop a full scope PSA are significant and could be amplified by MU considerations. A description of how the various phases of this project intend to address different MU risk aspects is described in Table VI–1.

This effort recognizes two general potential gaps in applying SUPSA methods and techniques to the MUPSA problem: (a) knowledge issues — limitations in the fundamental methods and data from SUPSAs when applied to MUPSAs; and (b) modelling mechanics issues — efficient means of building MU logic models, starting with SU models that are not cost-prohibitive. These are both important issues to address in order to formulate practical options without overly restrictive, overly complex approaches.

VI–2.3. Plants and PSA models considered in the EPRI project

The bulk of this effort was focused on the piloting of MUPSA considerations based on PSA models currently being used for risk informed activities by operating MU sites. This is based on the consideration that specific

225

TABLE VI–1. TOPICS FOR PHASES OF THE EPRI MU RISK
INVESTIGATION PROJECT

	MU topic	Phase 1	Phase 2	Phase 3
RIDM	MU RIDM framework	Initial framework	Refined framework	
State	Full power	Initial framework	Refined framework	
	Low power and shutdown			Initial methodology
Level	Level 1	Initial framework	Refined framework	
	Level 2			Initial methodology
	Level 3			Initial methodology
Hazards	Internal events	Initial framework	Refined framework	
	Internal flood	Initial investigation	Refined framework	
	Internal fire	Initial investigation	Initial framework	Refined framework
	Seismic	Initial investigation	Initial framework	Refined framework
	Other hazards	Initial investigation		
Method	HRA	Initial methodology	Refined methodology	
	CCF	Initial methodology	Refined methodology	

TABLE VI–1. TOPICS FOR PHASES OF THE EPRI MU RISK
INVESTIGATION PROJECT (cont.)

	MU topic	Phase 1	Phase 2	Phase 3
Other	Pilot plants	PWR dual-unit sites with internal events and internal flood SUPSAs	PWR dual-unit sites with internal flood, fire and seismic SUPSAs	MU sites with 2+ reactor units and other design aspects

modelling aspects and insights of MUPSA are very much plant specific and based on plant configuration on one side, but also modelling techniques, including aspects regarding PSA software and modelling philosophies, on the other. The three pilot sites investigated were selected with the intent of diversifying the possible challenges and insights, while maintaining enough similarities to make meaningful comparisons between the pilots. Two pilot sites are for plants operating in the USA, and one pilot site is for a plant operating in Europe, where:

(a)   All pilots are dual-unit sites with similar reactor designs on-site but with different configurations. Two pilots have shared auxiliary/control buildings, whereas the other pilot has essentially completely separated structures.

(b)   Two of the pilots use a single-top fault tree linking approach, whereas the other pilot uses an event tree linking approach. This has different implications for how the MU model development will progress.

(c)   Two of the pilots have explicit modelling of the individual units or duplicated the model to effectively obtain the same result, whereas the other pilot has an SU top that is used for both units with appropriate logical switches that modify unit specific configurations.

The three pilots have also different hazards and POSs explicitly modelled. One pilot has a PSA model that includes various aspects (albeit in different model versions, which represents another practical challenge for any risk aggregation activity), including internal flood and internal fire, LPSD and SFP PSA. One pilot has, beyond the internal events and internal flooding, a state of the art seismic PSA model. In the first phase of this pilot investigation, some basic considerations were made on the assembling of the MU logic and extension of CCF to additional units, focusing primarily on the internal events and on CDF results.

VI–2.4. Plant A

Pilot Plant A site consists of two four-loop PWR reactor units with ice condenser containments. The two units have independent reactor buildings but share a single auxiliary building (which includes the SFP), a single control building and two connected turbine buildings in a twin unit configuration. A diagram of the site layout is shown in Fig. VI–1.

The two units do not have significant cross-ties, but there is a significant degree of shared support systems. Most notably, the essential service water and the CCW systems are tightly coupled. The essential service water system provides the essential auxiliary support functions to the engineered safety features of the plant. The system is designed to provide a continuous flow of cooling water to those systems and components necessary to plant safety, either during normal operation or under accident conditions. The system consists of eight pumps, four travelling screens, four travelling screen wash pumps and four strainers located in the intake pumping station. Both trains of essential service water feed three trains of CCW systems, with the CCW system train B serving both units. CCW system 1B pump is normally aligned as a backup to train 1A (and the CCW system 2B pump is aligned for 2A). The CCW system train B can receive electric power from either train of power support. Power support is also partially shared but the Unit 1 shutdown boards (that is, engineered safety features switchgears) mostly support Unit 1, and vice versa for Unit 2.

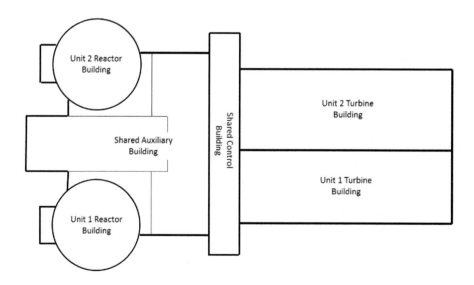

FIG. VI–1. General building arrangement for Pilot Plant A.

The circulating water pumping station is located in the yard between the turbine building and the cooling towers. There are eight circulating water pumps. Four pumps for each unit operate in parallel and circulate water from the cooling tower cold water basin, through the condenser and back into the heat exchanger section of the tower. The standby on-site power is supplied by four EDGs.

Major equipment location is generally arranged in a mirror image between the two units, with some notable exceptions such as the CCW pumps, which are all located in the Unit 1 side of the auxiliary building. Separate and similar safety related systems and equipment are provided for each unit, except as summarized in the list of shared equipment and/or components.

Pilot Plant A has a fully developed dual-unit at-power Level 1 and Level 2 PSA logic that addresses internal events and internal flooding in the same logic tree. Unit 1 and Unit 2 dedicated components are explicitly modelled in the PSA with a dedicated and unique BE nomenclature. This provides significant convenience when extending SU into MU risk assessment considerations with more direct PSA model changes. Shared components are indicated explicitly in the PSA model nomenclature. The internal events model addresses (a) unit specific initiators and (b) initiators that are already identified as impacting both units:

— Total LMF;
— LOOP (grid, plant and weather related);
— Total loss of plant service air;
— Total loss of essential service water;
— Partial loss of essential service water;
— Loss of component cooling system (partial or total);
— Loss of unit specific AC vital instrument board;
— Loss of unit specific vital battery board.

Despite being used and described primarily as SUPSA models, a significant amount of MU related aspects are indeed included via the IEs listed above (which have varying degrees of MU potential) and the systems and components that are considered explicitly as part of the PSA model. As in the other pilot plants, this allows for the consideration of how much of the existing model can be leveraged for MUPSA insights.

The internal flood assessment is performed on a room-by-room basis spanning the entire plant (that is, explicit flooding scenarios are present for mirror image rooms between the two units), generating unit specific impacts. The two units have a fairly symmetrical risk profile, with a CDF close to 9×10^{-6}/year. The comparison between the SUCDF results shows that a significant contribution of the SUCDF comes from initiators that are impacting both units, mainly as a result of internal flood scenarios. A non-trivial percentage of the remaining risk profile is also possibly impacted by MU considerations (e.g. loss of the CCW system or loss of vital boards for individual units due to cross-unit CCFs). The internal flood contribution is dominant to the SUCDF, mainly because of internal flood scenarios in shared locations of the building or scenarios that can propagate and impact both units.

VI–2.5. Plant B

The Pilot Plant B site has two four-loop PWRs with large dry containments, which are effectively mirror images of each other. A seismically qualified dam maintains a minimum water level for the UHS, sufficient to allow a simultaneous safe shutdown of both units (assuming one unit is in a LOCA) for a minimum of 30 days without make-up. All four service water pumps (Units 1 and 2) are located in the service water intake structure, and all four pumps are normally operating. Eight circulating water (that is, normal service water) pumps (four for each unit) are also housed in a single circulating water intake structure.

There are several shared systems between the units and several capabilities to cross-tie functions. Regarding the cross-ties, the service water and CCW systems are credited for limited scenarios in the model. By procedure, the supplying unit will sacrifice one train given a total loss in the other unit. These cross-ties are very important to this analysis because they provide (a) significant means by which the benefits of being a dual-unit site might overcome the added risks and (b) a significant area of investigation for site wide risk, including the potential for cross-tie failure to induce a dual-unit event.

With regard to operation, the two units operate under the same procedures, and operators are licensed and routinely assigned to both units in a rotation. The outage schedule for the two units is staggered as much as possible for 18 month fuel cycles, while avoiding summer outages (typically spring and autumn outages, with an occasional dual-outage year). Outage duration is normally about three weeks in length with about a five day offload window. Work on safety systems is done on a train basis so that key safety function defence in depth is maintained. In general, both emergency core cooling system trains are maintained until the cavity is flooded. After flooding, one train is always maintained with alternative functional paths.

The Pilot Plant B PSA model is an SU model that can be aligned to quantify either Unit 1 or Unit 2. Significant differences between the units are explicitly modelled with the use of house events (for example, switchyard alignment and tube rupture, because one unit has a newer steam generator). However, there are no significant differences in numeric results or risk insights between the units. The Unit 1 alignment is therefore typically used as the surrogate for either unit.

The Pilot Plant B PSA model includes internal events, internal flood and internal fire PSA models. The internal events and internal flood PSAs are used in the MUPSA investigation. The internal fire PSA is currently being leveraged for MU risk purposes under this effort. The PSA documentation was reviewed to identify shared and cross-tied systems. The following list summarizes these:

— Instrument air common compressors;
— Fire protection system;
— Service water intake structure;
— Uninterruptible power supply heating, ventilation and air-conditioning;
— Control room heating, ventilation and air-conditioning;
— CCW system;
— Safety chilled water;
— Station service water;
— 6.9 kV and DC electrical;
— Common buses;
— Turbine plant cooling water;
— SFP cooling.

The baseline model for Pilot Plant B already includes many dual-unit issues in its SU model, such as LOOP affecting both units (e.g. disallowing credit for other unit equipment), dual-unit CCF for station service water, CCW, safety chilled water and emergency power systems, as well as probabilities for the other unit being unavailable because of a concurrent initiator that would impact on common and/or shared functions. Pilot Plant B also has an alternative power EDG for each unit that acts as a backup power source if the main EDGs fail. The CDF results (including internal events and internal flooding) for Unit 1 and Unit 2 are approximately 4×10^{-6}/year (where disabling all credit for cross-ties increases the CDF to 1.7×10^{-5}/year).

The PSA includes a simplified Level 2 model, using detailed plant specific evaluations for every input. The LERF PSA model is a subset of the Level 2 model and is approximately an order of magnitude lower than CDF results, which is fairly typical for PWRs with large, dry containments. The LERF is primarily due to early containment bypass events.

Although the plant has fewer fire compartments compared with other plants, the compartments are generally larger than in similar, typical plants. The large volume of the fire compartments results in a low frequency of scenarios capable of generating damaging hot gas layer conditions. The most significant contributing risk factor to the fire PSA is abandonment of the MCR, either from loss of functionality or loss of habitability.

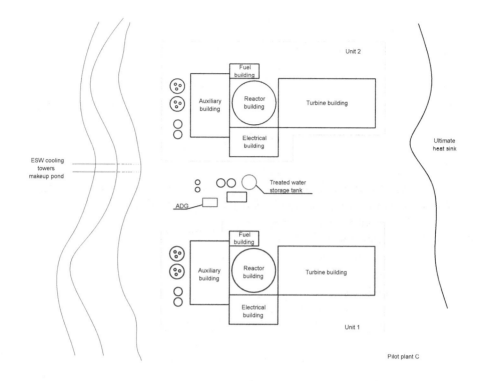

FIG. VI–2. Pilot Plant C site overview.

VI–2.6. Plant C

Pilot Plant C is a dual-unit site consisting of two units of three-loop PWRs with large dry containments (see Fig. VI–2 for a site overview). The two units have practically the same characteristics. The units are functionally and physically separated, sharing few SSCs.

The relevant differences between the units from a PSA logic model point of view were assessed as part of the baseline SUPSA, with the conclusion that these differences are not significant, as follows:

(a) All design changes considered in the current revision of the PSA model have been implemented in both units, with the same as-built, as-operated configuration.

(b) No differences have been identified in the frontline and support systems credited in the PSA; therefore, the systems analysis is considered applicable as-is to Unit 1 and Unit 2. While minor component and cable tag number differences between the units have been found, no impact on the analysis is contemplated as equivalent components exist.

(c)   The data analysis already includes equipment unavailability and failures of both units. Moreover, insights from the operations staff indicate that the behaviour of both units regarding equipment unavailability and failures is similar.

(d)   Operating procedures are identical in both units (e.g. emergency operating procedures). Operators' training is developed under an almost identical structural policy in both units. Similarly, this is also applicable to the procedures for testing, maintenance, calibration and realignment.

The units' cooling system consists of a natural draft cooling tower, two forced-draft cooling tower sets and the UHS. During normal operation, the CCW system transfers the thermal load from safety related equipment to the UHS through the normal service water system. During emergency operation, the thermal loads are transferred to the essential service water system. The heat is transferred to the atmosphere through the forced-draft cooling tower sets. The essential service water system has two identical and independent trains.

The unit-dedicated systems of each unit are identical and do not possess cross-tie capabilities. Unit 1 systems are fully independent from those of Unit 2 and vice versa. With few exceptions, the units share almost no safety related SSCs. The only shared or cross-tied SSCs credited in the PSA model are the following:

(a)   A safety (emergency) water supply from the essential service water system cooling features.

(b)   A non-safety (normal) water supply from the treated water storage tank.

(c)   Alternative diesel generator (ADG) which can be started and aligned to feed one AC busbar train of each unit (the ADG capacity can only feed one unit). Two common breakers (each connecting to the corresponding unit) have an interlock that prevents the inadvertent connection of the ADG to both units, which could result in overloading and tripping the ADG.

The Pilot Plant C PSA model includes only the modelling for one unit (Unit 1). The PSA model only has Level 1 internal events at power, although minor portions can be used for Level 1 LPSD purposes. Limited cross-ties between units exist. The 'cross-train' alignment operation, where the charging pump is aligned to CCW system train B and the reactor coolant pressure thermal barrier is aligned to CCW system train A, helps decrease the risk, and it is the current plant's operational practice. Not crediting this cross-train alignment would increase the overall CDF by about 64%. The ADG is credited in the SU model, fully dedicated to the unit under study. The ADG unavailability due to its potential use by the other unit had not been considered in the baseline SU model. The only equipment modelled in the baseline PSA model that is common to both units is the essential service water cooling tower make-up system (normal and auxiliary means) and the ADG. Pilot Plant C internal events at-power CDF results for the Unit 1 (also applicable to Unit 2) baseline model calculate a CDF of approximately 6×10^{-6}/year.

VI–2.7. MUPSA methodology

As stated above, the first phase of this effort focuses on the maximum leveraging of the comprehensive PSA-focused approach to deriving risk insights for aspects specific to multiple units on a plant site. Posterior phases will focus on expanding and validating the second part of the framework (e.g. potential simplified MU risk assessment approach options).

VI–2.7.1. Leveraging risk insights from plant information and SUPSA models

The first task is to review existing site characterization, plant information and the SUPSA model(s) to identify what is relevant (and not relevant) in available information with regard to MU risk. This can both bound the scope of the MUPSA development (rather than develop all new information) and provide insights before any risk assessment is performed (e.g. which hazards and systems are more relevant for MU aspects). This is particularly useful for regulatory regimes without explicit MU risk requirements, since the insights already exist and can be leveraged for plant safety aspects. Leveraging both plant information and SUPSA models can confirm specific MU insights, as well as provide an approximate understanding of risk contribution (some of which may need to be confirmed quantitatively in a MUPSA modelling effort).

Sites are expected to have some variability with regard to the overall importance of MU risk; that is, the degree to which MU risk will be significant compared with SU risk and the degree to which MU risk insights might be different from SU risk insights. For some sites, the MU risk may be a small contribution to overall site risk, whereas for other sites, MU risk may be a dominant contributor. The approach to evaluating MU risk needs to be 'risk informed' in the sense that the resources used to assess the risk have to be commensurate with the initial assessment of MU risk based on the site characterization and qualitative considerations.

Identifying plant features related to MU risk that have common attributes across units is the next key aspect. Important attributes include physical location (common grid, common and unit specific capabilities), structures (shared, connected), systems (shared, cross-tied), components (shared, identical), resources (shared physical, human, procedural) and command and control (emergency organization).

A key aspect in reviewing plant information is to understand the level of coupling between the units. Most reactors co-located within a site will exhibit some form of coupling (i.e. distances between units will be relatively small when compared to multiple sites). However, as confirmed by the pilot plants, some units will have a varying degree of coupling in terms of shared structures and components.

In parallel or in series to the above, the SUPSA model and documentation to identify the MU risk elements need to be considered, including the extent to which the SUPSA model may need to be changed to address MU risk. The SUPSA may have included some elements important to MU risk, such as shared systems, internal flood progression from one unit to another in shared structures, common fire scenarios in connected structures and seismic fragilities for individual unit components. The SUPSA may also have identified differences among units that would be helpful in justifying limited dependency modelling in some areas. It is important to understand the basis for the SU model because it may be based on a limited SUPSA model or the bounding characteristics of the units. Also, the SUPSA model may include primarily the positive contribution of shared systems or cross-ties from the unaffected unit to the affected unit.

The SUPSA may have included some elements important to MU risk, such as shared systems, internal flood progression from one unit to another in shared structures, common fire scenarios in connected structures and seismic fragilities for individual unit components. The SUPSA may also have identified differences among units that would be helpful in justifying limited dependency modelling in some areas. It is important to understand the basis for the SU model because it may be based on a limited SUPSA model or the bounding characteristics of the units. Also, the SUPSA model may include primarily the positive contribution of shared systems or cross-ties from the unaffected unit to the affected unit.

VI–2.8. Leveraging operating data for MU risk purposes

Gathering data relevant for MUPSA modelling can represent a significant cost for addressing MU risk. Hence, it is critical to understand how the key sources of data can be gathered, determine how to leverage the data for potentially screening lower or non-contributors and implement this within the existing PSA framework.

Similar to the traditional concept of POSs, the extension of MU aspects can consider SOSs, where a variety of combinations in MU sites can be considered. For two-unit sites, this is relatively straightforward at a high level: three basic SOSs can be defined, depending on which unit is at full power and/or which unit is in a shutdown condition. For convention purposes, this effort uses the following nomenclature (recognizing that this could be done in a number of ways): SOS-22 for 2-of-2 units operating, SOS-12 for 1-of-2 units operating, regardless of which unit, and SOS-02 when both units shut down. As noted previously, the initial phases of this effort focus on SOS-22, i.e. both units operating (with future phases intended to investigate combinations in further detail).

Data can be gathered to represent the fraction of time that the site is in each SOS, based on a long term average (these are preferably calculated from site specific experience, if available). In this effort, these fractions can be calculated using existing plant availability factors for each unit and one new parameter, the site availability factor, which represents the fraction of time that both units were operating.

Data maintained by the NRC [VI–25] were used to derive insights into how this data may be developed and used in the pilot plant effort, based on US information. Table VI–2 shows the SOS fractions for five dual-unit sites to illustrate the range of fractions. Table VI–2 shows that SOS-22 dominates, with a narrow range from ~80% to 85%. The SOS-02 results show a much broader range, from ~1% to 0.05% (either way, a very small fraction for all sample plants used).

Table VI–3 shows the results for three-unit sites where, again, the dominant state (from an SOS fraction standpoint) is all units operating (SOS-33, in this case) with a smaller but significant contribution from SOS-23, where one unit is shut down. In line with focusing on the more significant contributors, as well as bounding the scope of the MUPSA modelling, SOS-13 and SOS-03 might be screened based on their low fractional occurrence rate, with a review of the conditions in the specific outages in these states to verify no significant risk contributions. Note also that state SOS-03 might result from site wide IEs (e.g. LOOP), in which case the initial 24 hours of that state may be captured by the initiator with the plant in state SOS-33. Therefore, the insight is not the need to consider three units in refuelling outage concurrently but that they should be considered in MUIE frequency estimation.

For sites larger than three units, the SOS combinations become more complex, although operating experience may indicate well-defined patterns. It

TABLE VI–2. SAMPLE CALCULATIONS FOR TWO-UNIT SOS FRACTIONS (FOR ILLUSTRATION PURPOSES)

SOS	Plant 1	Plant 2	Plant 3	Plant 4	Plant 5
SOS-22	85.5%	83.2%	82.7%	80.2%	84.4%
SOS-12	14.4%	16.3%	17.0%	19.6%	14.6%
SOS-02	0.05%	0.5%	0.3%	0.2%	1.0%

TABLE VI–3. SAMPLE CALCULATIONS FOR THREE-UNIT SOS
FRACTIONS (FOR ILLUSTRATION PURPOSES)

SOS	Plant 1	Plant 2	Plant 3
SOS-33	70%	82%	73%
SOS-23	28%	17%	26%
SOS-13	2%	0.3%	1%
SOS-03	0.3%	0.3%	0%

may also identify dependencies between subsets of units on a site if some groups
of units are more likely to be in a shutdown state than other units on a site.
The insights from this data analysis clearly indicate that this step is critical in
understanding the general scope and determining the need for a detailed analysis
for a large number of SOSs. For a site that is either required to perform detailed
MUPSA modelling or voluntarily chooses to do so, this data analysis does not
prevent quantification of multiple SOS combinations. However, even a cursory
look at the data above suggests that focusing resources on the most significant
fractions is reasonable.

VI–2.9. Collecting data for MUIE frequency characterization

For the purposes of developing the necessary frequencies to characterize
IEs impacting units at a site level (as well as potential cascading events), specific
scenarios and hazards need to be considered. In line with the above discussion,
the extent and scope of the IE frequency analysis has to be performed judiciously.
Not all IEs may be significant contributors, whereas some (e.g. site wide LOOP
with or without external hazards) will be expected modelled contributors.

A note should be made about so-called 'cascading' IEs. These are typically
considered IEs that start in one unit and lead to an initiator in other units (which
would not have occurred in the absence of the first unit impact). It is possible
to envision two loose types of cascading IEs: one with couplings between units
due to dependencies, and another where completely independent events occur in
multiple units within the same, narrow time window. Similar to component CCF,
the independent likelihood of phenomenologically distinct initiators (e.g. LOOP
and LOCA) is expected to be small (to the point of screening a large combination
of such independent events). The remaining type (which may or may not be
defined as cascading) need to be considered in the MU risk process, as long as

there is a compelling dependency that requires understanding of its potential overall contribution. This is both a technical and practical consideration; for example, calculating a large combination of events that are extremely unlikely to occur together will consume significant resources but will not provide additional benefits or risk insights. This also places more responsibility on properly and accurately capturing potential dependencies between units to avoid overlooking potential contributors.

An analysis of MUIEs was performed using a well-maintained database that includes international data [VI–26]. A data set of more than 1600 reactor trip events over a ten year period around the world was collected. These events are mostly plant transients that include an automatic or manual reactor trip, occurring either as SU events or MU events, but include LOOP initiators (see Table VI–4 for further details).

After parsing the data, it was found that a total of 1422 trips occurred on MU sites, including 1341 SU trips and 81 MU trips. These 81 MU trips were due to a range of 35 site initiators, where site-initiator events included reactor trips that occurred at multiple units (two or more) on the same site.

Hence, at MU sites, about 6% of all trips are MU trips (81 out of 1422). Therefore, the vast majority (94%) of the trips in MU sites are SU trips. This

TABLE VI–4. DATA ANALYSIS — REACTOR TRIP RATE FOR SU AND MU SITES

Sites	SU trips	MU trips[a]	Operational data	SU trip frequency	MU trip frequency	Total trip frequency
MU site	1341	81	3482.7	0.385	0.023	0.408
SU site	244	0	438.1	0.557	n.a.[b]	0.557
Total	1585	81	3920.8	0.404	n.a.	0.425
Units	Number of SU reactor trips	Number of unit reactor trips involving MUs	Reactor operating-years	Events per reactor operating-year	Events per reactor operating-year	Events per reactor operating-year

[a] The count of MU trips includes each unit that tripped. MU trips are reactor trips that occurred at units on the same site on the same day. Therefore, for example, a site level initiator where all three units tripped is counted as three MU trips.

[b] n.a.: not applicable.

238

factor, expressed as a fraction (81/1422 = 0.0570), is the conditional probability of an MU event, given a reactor trip event has occurred on an MU site. Obviously, this factor applies only to MU sites, because SU sites cannot have MU events.

The data show that the average annual reactor trip rate for MU sites (0.408) is similar to but somewhat lower than for SU sites (0.557) — about one trip every two reactor operating-years. MU sites may have a lower trip rate because of shared systems that allow units to remain at power for conditions that would force an SU site to shut down (reinforcing the benefits of the ability to share and/or cross-tie systems and components). However, the annual MU trip rate (0.023) is much lower than the SU trip rate, either at SU sites (0.557) or MU sites (0.385), by a factor of 20. The MU trip rate represents an average of one MU trip every 45 reactor operating-years.

This is an area where consistency of the use of the data is critical. These results are trips per reactor unit, not per site, because trips occur to reactors, not to sites. The frequency units are per reactor operating-year. However, the MU trip rate (0.023) is on a per unit basis but represents an MU event, that is, a reactor trip of at least two units on an MU site. The 81 MU events were reviewed and classified in one of seven groups, as follows:

(1)   MU trip, LOOPG — multiple unit trips due to grid related LOOP;
(2)   MU trip, LOOPX — multiple unit trips due to weather (or earthquake) LOOP;
(3)   MU trip, partial LOOP — multiple unit trips involving partial LOOP;
(4)   MU trip, UHS challenge — multiple unit trips with challenge to common UHS;
(5)   MU trip, cascading — multiple unit trips, where an SU event impacts the second unit;
(6)   MU trip, manual — multiple unit manual reactor trips;
(7)   MU trip, independent — multiple units trip (same site and day), but from independent causes.

Only two of the 35 MU trip events were considered cascading events, where the response to an SU trip led to a second unit trip. Three MU trip events were involved multiple SU trips on the same day at the same site. Of the 1666 reactor trip events that occurred internationally in this ten year period, 172 reactor trip events (10.3%) were due to external hazards. Of these, only 38 (22.1%) were MU trips. Note that, the count is based on the number of unit trips. These 38 unit trips occurred because of 16 site level external events. Also note that this counts only NPPs that were operating; units that were shut down would not have recorded a reactor trip.

This analysis of a large set of international events allows for some broad conclusions and recommendations regarding MUIEs. Because of large site to site variabilities, these data are not intended to provide generic frequency estimates. Instead, they are used to provide estimates of the fraction of existing, site specific initiator frequencies that have to be assigned to the MU initiators. Limited availability of data means the generic fraction would be the initial estimate for a site specific analysis, but that fraction might be adjusted based on site specific characteristics. Initial estimates are shown in Table VI–5.

VI–2.10. Assessing CCF data/modelling for MU aspects

The practical purpose of the treatment of CCF in this effort is to explore the basis for CCF parameters in the context of MU risk assessments. Initial work is described in this area, which is currently being expanded for this project.

TABLE VI–5. FREQUENCIES OF MU EVENTS BY CATEGORIES, 2008–2017

Category	Description	Event frequency[a]
MU trip, LOOPG	Multiple unit trips due to grid related LOOP	4.88E-03
MU trip, LOOPX	Multiple unit trips due to weather or earthquake LOOP	4.02E-03
MU trip, partial LOOP	Multiple unit trips involving partial LOOP	7.18E-03
MU trip, UHS challenge	MU reactor trips with challenge to common UHS[b]	3.16E-03
MU trip, cascading	Multiple unit trips, where SU event impacts second unit	1.15E-03
MU trip, manual	MU manual reactor trips	1.15E-03
MU trip, independent	Multiple units trip on same site and same day, but from independent causes	1.72E-03

[a] The event frequency is specified in events per reactor operating-year for units on MU sites.
[b] UHS was not lost for any of these events.

The process of limiting CCF groups is an important step in addressing model mechanics, i.e. making MUPSA models that are solvable but also adequate to identify risk insights. Note that in the pilot plant studies performed for this first phase of the MUPSA project, CCF modelling was performed conservatively. It was assumed that all CCF events that form the basis for CCF parameters were lethal failures, i.e. if two out of two components in the group failed, all components would also have failed if the group size were expanded. This provided a simple means of expanding CCF modelling without extensive expansion of the underlying PSA model with large numbers of cut sets to include CCF contribution from subgroups. However, it is also understood to be a very conservative, bounding model because most of the CCF events available in existing databases are non-lethal failures.

One of the sources of generic CCF parameters is from the NRC CCF Parameter Estimations [VI–27]. This database provides CCF parameter estimates for a number of components and failure modes based on data from NPPs from 1997 to 2015. These data consist of 373 CCF events (out of 7666 independent failure events). Of these events, only four are categorized as 'lethal' failure events, while the remaining 369 are non-lethal. This issue is relevant when using CCF events to generate CCF parameters of larger groups than the original data. Extending the use of CCF to larger groups can become an issue because most CCF data involve two components only.

Table VI–6 provides the CCF parameters for a sample list of components that may be important in a MUPSA model. This table shows the count of CCF events and independent events for each component failure mode. It also provides the CCF parameters for CCCG of size 2 and size 4 for comparison.

The $\alpha2$ parameter estimates the fraction of failures that are in a group of two. For an SU CCF model, the common cause group CCCG-2 is expressed as (A1B1), where A1 and B1 represent the two redundant components in an SU common cause group. For a two-unit CCF model, this group is expanded to include the identical components in the second unit, that is, CCCG-4 (A1B1A2B2). Then, the A1B1 combination (for SU) now expands to four combinations for the two-unit model: A1B1, A1B1A2, A1B1B2, A1B1A2B2. Then the $\alpha2$ for A1B2 from the CCCG-2 ($\alpha22$) can be compared with the parameters from the CCCG-4: $\alpha24 + \alpha34 + \alpha34 + \alpha44$, because both expressions represent the conditional probability of at least two component failures. Because of the nature of the parametric approach to modelling CCF and the sparseness of the data, the extrapolation of CCF estimates to larger common cause groups may become overly conservative. Using the data shown in Table VI–6, a case when the Unit 1 components (A1, B1) have more combinations of CCF failures in a two-unit situation, the probability of A1 and B1 failing could be estimated as being higher than the SU case.

However, these components' failure modes have few CCF events in the data set considered (from zero to six). A list of the 23 actual CCF events is found in the CCF event count in Table VI–6. Most of the actual CCF events are 2-of-2 failures (two failures in a group of 2 or 2|2), 2-of-3 (2|3) or 2-of-4 (2|4). For the majority of the entries provided in typical CCF factor estimates, the CCF events have weighting factors (e.g. coupling strength or time delay factor) of less than 1.0 (that is, most events are not actual CCF events but marginal CCF events). Some of the other entries also appear to be marginal CCF events based on the event description. As a result, the α3 and α4 parameters depend strongly on the parametric process and the scarcity of data (see appendix C.4 of NUREG/CR-5485 [VI–28]) applied to a limited number of marginal CCF events (in other words, how to account for marginal CCFs being scaled to be included in CCF estimates for use in PSA models).

The bottom line of the CCF estimate characterization for MU risk aspects is that (a) data on which to base the parametrization process will be even more scarce for evidence of actual MU CCF events, and (b) how the accounting, weighing and characterization of the CCF data and estimates is performed can significantly drive the results (most likely, artificially). Further indication of CCF as a top contributor for MU issues necessitates careful consideration of its probability values, relevance to actual data and potentially significant uncertainty. This does not imply that CCF is not critical to proper risk modelling or should not be used. Instead, it highlights the need for careful assessment of the overall results. In addition, the cost of CCF modelling can increase exponentially, so practical issues (such as screening) need to be considered.

VI–2.11. Issues related to HRA for MU aspects

There are clear potential distinctions in operator responses and the reliability of operator actions given a MU accident, in contrast to an SU accident. However, the accident sequences in response to a MU initiator are still most appropriately considered as multiple SU accidents, particularly when each unit has a dedicated control room, control room crew and unit specific emergency operating procedures and abnormal operating procedures. MU accidents will still rely on standard HRA approaches that include identification of actions that need to be considered in response to an accident sequence, definition of the action in terms of procedural guidance and the accident sequence context and quantification of the HEP based on consideration of performance influencing factors[1]. However, the accident progressions of each unit on the MU site may have cross-unit dependencies due to site level initiators, shared SSC and shared

---

[1] Sometimes referred to as performance shaping factors.

TABLE VI–6. CCF PARAMETERS FOR A SAMPLE LIST OF COMPONENTS RELEVANT TO MUPSA MODELLING

Component type	Failure mode	No. CCF, independent events [a]	CCCG size [b]	Alpha factors for CCCG			
				A_1	A_2	A_3	A_4
Auxiliary feedwater pump, turbine driven	Fail to start	0, 89.0	2	0.996	4.4E-3	—[c]	—
			4	0.986	9.0E-3	3.0E-3	1.6E-3
	Fail to run	0, 59.3	2	0.994	6.2E-3	—	—
			4	0.983	1.1E-2	3.8E-3	2.1E-3
Auxiliary feedwater pump, motor driven	Fail to start	2, 44.7	2	0.960	4.0E-2	—	—
			4	0.965	1.6E-2	1.1E-2	6.9E-3
	Fail to run	2, 18.2	2	0.973	2.7E-2	—	—
			4	0.956	3.5E-2	5.7E-3	2.9E-3
CCW pump, motor driven	Fail to start	0, 92.7	2	0.996	4.2E-3	—	—
			4	0.987	8.7E-3	2.9E-3	1.6E-3
	Fail to run	1, 72.1	2	0.989	1.1E-2	—	—
			4	0.981	1.4E-2	3.3E-3	1.8E-3
Service water pump, motor driven	Fail to start	0, 30.5	2	0.989	1.1E-2	—	—
			4	0.976	1.6E-2	5.2E-3	2.8E-3

TABLE VI–6. CCF PARAMETERS FOR A SAMPLE LIST OF
COMPONENTS RELEVANT TO MUPSA MODELLING (cont.)

Component type	Failure mode	No. CCF, independent events[a]	CCCG size[b]	Alpha factors for CCCG			
				A_1	A_2	A_3	A_4
	Fail to run	3, 71.0	2	0.971	2.9E-2	—	—
			4	0.970	1.7E-2	7.8E-3	4.9E-3
Emergency generator diesel	Fail to start	4, 261.9	2	0.991	9.5E-3	—	—
			4	0.992	4.6E-3	2.5E-3	1.2E-3
	Fail to load	5, 275.3	2	0.994	5.7E-3	—	—
			4	0.987	1.1E-2	1.4E-3	0.7E-3
	Fail to run	6, 217.1	2	0.984	1.6E-2	—	—
			4	0.986	6.7E-3	5.1E-3	1.9E-3

[a] No. CCF, independent events: number of CCF events and independent events for the component type and failure mode.
[b] CCCG size: size of the common cause component group, either 2 or 4.
[c] n.a.: not applicable.

human and organizational resources. The definition and quantification of some actions may need to be adjusted based on the potential dependencies across units.

Two potential differences between SUPSAs and MUPSAs have to be considered: (1) If the units on the site are significantly different in design and/or operation, the SUPSAs need to be reviewed to assure that any unit specific differences in required operator actions have been identified; and (2) the MUPSA may identify additional actions important to MU risk that were not modelled in SUPSAs. These actions would need to be defined and quantified using the standard HRA methods. However, HFE definitions may need to be modified for several types of operator actions to reflect the potential performance influencing factors that may be affected by dependencies among units, as follows:

(a) Operator actions related to shared or cross-tied systems; potentially more complex with regard to cognitive decision making/execution, and may involve a wider set of site resources with different timing for responses and realignment impacts;

(b) Operator actions allocating shared components or resources, especially if there is a need to decide how limited resources are used, that is, which unit to prioritize by allocating shared components or resources;

(c) Operator actions outside of the MCR; for local actions, the severe accident progression for MUs may impact accessibility to locations where actions are performed.

The quantification of HFEs for MU accidents needs to account for the differences in HFE definitions discussed previously. However, the quantification also needs to address the potential for implicit dependencies among units based on the same accident sequences. Even if the units are operating essentially independently (that is, without some of the explicit dependencies described previously), some level of cross-unit dependencies may be appropriate, based on the degree to which the plant context is the same for each unit (e.g. same procedures, training, experience; same instrumentation, cues, human–machine interface; similar sequence timing, response timing).

These dependencies will need to be treated implicitly because the current HRA methods do not identify the specific cause of the HFE. As a result, the dependencies are based on a shared 'plant context', but it is difficult to identify explicitly how important each aspect of the plant context is in estimating the reliability of the operator action. The specific level of dependency (zero dependency to complete dependency) has a large degree of uncertainty, but the relative dependency levels can be justified based on the units' similarity (that is, the more similar, the higher the dependency level) and on the action timing (that is, short term actions have a higher dependency level, with the exception of initial actions).

HRA approaches for MUPSAs will be refined further in the next phase of this project, including insights from operator interviews and recent publications [VI–29].

VI–2.12. MU risk modelling and insights from pilot plant studies

The approach (and data) discussed in previous sections was implemented for the three pilot plants mentioned before (Plants A, B and C). Full details are available in an EPRI report [VI–24]; a brief summary is provided here.

Because of the explicit modelling of the individual unit components and associated failure modes for both units in the PSA model for Pilot Plant A, the development of the MU logic was relatively straightforward for this pilot. The MUPSA logic was developed using a simple top logic approach, as illustrated in Fig. VI–3, and quantified for the Unit 1 top gate, the Unit 2 top gate and the Unit 1 AND Unit 2 top gate (that is, gate 'U12_CDF'). With additional logic structures, the model could be used to quantify individual unit CDF results, as well as MU risk contributions, along with the inclusion of plant availability fractions for proper representation of the results.

Only the LOOP initiators, the complete loss of the emergency service water initiator and the loss of the plant air (fully shared) system are modelled as MU initiators. The vast majority of the internal flooding initiators are also modelled as

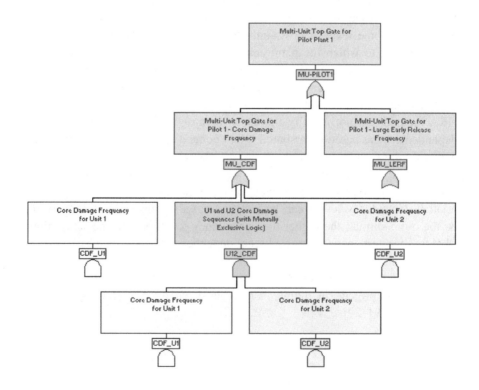

FIG. VI–3. Top logic for Pilot Plant A, MUPSA.

MU initiators (some exceptions remain for flood scenarios that are only expected to impact one unit). In addition, the bulk of independent SU initiators are not expected to impact the second unit. Fully independent concurrent initiators have a combined IE frequency that would result in very low risk significance already, especially when considering the IE frequency and the narrow mission time.

Each unit includes a partial LMF IE and also a total LMF IE. It can be argued that a total LMF has significant potential to be due to a CCF event, especially in a tightly coupled plant like Pilot Plant A, which has two essentially identical units. The IE frequency associated with this initiator is also relatively high (that is, approximately 1×10^{-1}/year). A sensitivity case was therefore designed for which the partial LMF events remain unit specific, while the total loss is assumed to be a dual-unit event.

The loss of the essential service water system is modelled through IE fault tree logic for an SU (that is, partial) loss of the system and then for a dual-unit total loss of the system. In this case, there is no need to model additional MU initiators (that is, two partial losses of the system) because a double partial event is essentially included in the total loss of the system event.

The loss of CCW system is modelled through IE fault tree logic. For each unit, two initiators are modelled: a partial loss and a total loss of CCW system. Because train B is shared between the units, a total loss of CCW system in one unit also impacts the second unit, and it is therefore appropriate to model it as a potential MU event.

CCF BEs associated with Unit 1 and Unit 2 components have been modelled in a simplified and conservative fashion in the extension of the SUPSAs to MU metrics. Namely, all of the CCF events for each unit have been reviewed, and the lethal event in each group has been assumed to be a lethal event across the units. For example, both units include a common cause group associated with a failure mode for steam relief valves. The lethal CCFs for both groups were made a MU CCF group.

Differently from the individual component failure modes, the naming convention for operator actions does not differentiate between units in Pilot Plant A. No changes were made to the HRA naming convention, which would mean that fully correlated actions between units are envisioned in the current version of the Pilot A MUPSA logic, as an initial conservative assumption. Table VI–7 summarizes the results of the base case MUPSA estimates for Pilot Plant A.

The changes in the SUCDF values are not significant, and do not significantly change with or without flooding. A slight increase in the SUCDF can be expected because of the reduced reliance on the other unit. The contributions of initiators associated with the MUCDF with and without flooding are approximately 4×10^{-6}/year and 2×10^{-7}/year, respectively. The dominance of the internal flood scenarios in the MU contribution is not per se a unique insight

TABLE VI–7. PILOT PLANT A RESULTS FROM THE MU MODEL

UNIT	Internal events and internal flood		Internal events only	
	CDF from MU model	Change from base case	CDF from MU model	Change from base case
Unit 1 (CDF_U1)	9.11E-06	0.37% increase	5.13E-06	0.63% increase
Unit 2 (CDF_U2)	9.01E-06	0.26% increase	5.01E-06	0.47% increase
Two-unit CDF (U12_CDF)	3.84E-06	n.a.[a]	1.96E-07	n.a.

[a] n.a.: not applicable. The base case model does not have a MUCDF end state and, therefore, the MUCDF from the MU model cannot be compared to the other risk metrics.

derived from the MUPSA work. The internal flood risk is a known risk scenario at the site, which is considering adding dedicated flood alarms in proximity of the flood sources. When looking at these scenarios in a MU perspective, some additional considerations can be made. The SU model currently assumes that all of these flood scenarios are impacting both units, mainly because of the impact on critical electrical equipment, therefore causing a trip of both units. This modelling does not take into consideration any timing difference in the generation of a unit specific initiator because, after the propagation path is completed, both units are impacted. In reality, it is to be expected that one unit will trip before the other, because of water impacting electrical equipment in that unit before further propagating and equally impacting the other unit. As it is expected that operators would be dispatched to the plant after the trip of the first unit, positive identification of the event may be reached before the second unit trips. When the plant specific impacts associated with internal flooding scenarios have been eliminated, LOOP is the critical initiator from an MU perspective, followed by the fully shared emergency service water system and then from a combined loss of the CCW system, which shares only a train between units (but now has a CCF BE for dual-unit impacts).

A review of importance measures for the base case with internal flooding is minimally informative because of the overwhelming importance of the internal flood initiators, since many are assumed to result in direct or almost direct core damage scenarios. The CCFs between independent systems across units do not have a significant impact on the MU results and, therefore, appear to indicate that the need for explicit and accurate MUCDF modelling is minimal.

Phenomenological elements between units (that is, dependencies and correlation) are extremely important in the modelling of MU aspects, as expected. Full correlation of such events is a preliminary conservative approach that can reduce the necessary scope of the MU modelling, as well as highlight areas of more risk significance. However, this can also have the effect of masking scenarios where the units are proceeding to core damage following different sequences, which would allow for unit specific mitigation strategies (that is, by accounting for different timing between the sequences).

Pilot Plant A has a fully developed seismic PRA model, capable of quantifying unit specific seismically induced CDF and LERF. Dedicated fragilities are performed for the two units, and the model is quantified at eight different seismic hazard intervals. A MU version of the seismic PRA was developed in a manner similar to that applied for the internal events MU model. Because of the limited impact of the CCFs that have been added or modified for the MU model, the MU seismic PRA was not modified in the same manner (that is, MU CCF contributors were not added). This is assumed to be a reasonable assumption, because the fragility groups retain the same names when mapped to Unit 1 and Unit 2 components, and the correlation of seismic failures between units is preserved and is expected to dominate the results. As is commonly done, the quantification of the seismic PRA was performed for the individual units and corrected for the rare events approximation failure at higher g levels with a cut set post-processing tool that quantifies the top cut sets with a binary decision diagram approach. This corrects the overestimation of seismic CDF in the SU models, which is already relatively limited. The baseline seismic CDF for the SU models is approximately 2×10^{-6}/year and is similar for both units.

For Pilot Plant B, an initial MU model is built from the existing SU model through the identification of MUIEs and characterization of the various SOSs. Each SOS is modelled using the applicable portions of the SU model, and additional logic is used to ensure that each state can generate MCSs and to integrate dependencies. For this method, post-processing the cut sets plays a significant role. The initial model development is then modified as needed to address the details of various dependencies, such as MUIEs, common and/or shared equipment and impacts to CCFs and HRA.

Because the baseline is an SU model that can quantify for either unit, the general idea is to use the existing cut set frequencies to represent one unit (analysed unit) with concurrent probabilities that the other unit experienced an event and/or core damage (that is, CCDP) and to set up the dual-unit model to allow for as much visibility in cut set-space as possible. To describe this in another way, for each cut set from one unit, every cut set from the other unit is potentially possible; and every cut set from the other unit, for a given IE, is equal to CCDP. The initial step is to set up the framework for the cut sets and then to modify or remove CCDP events based on the events that led to core damage.

This approach avoids the need for duplicating the model, but it admittedly creates some difficulties in generating accurate importance measures, and it requires careful treatment and interpretation of the results. However, it is not expected that this loss in accuracy of the importance measures would significantly impact risk insights. Most other issues that were encountered with this approach are common to the issues of using other approaches (that is, the need to identify and account for the details of MU dependencies). Full details on how this was developed and implemented are discussed in an EPRI report [VI–24]. A detailed screening of CCF BEs was performed, similar to Plant A.

The dual-unit results for Pilot Plant B (see Table VI–8) are heavily dominated by LOOP. While LOOP clearly dominates the results, other potentially MU events of potential risk significance are identified. Loss of service water (MU event) appears in the combined results, but because the overall dual-unit contribution is only about 3% of the total risk, its overall contribution competes with SU events. Loss of CCW does not appear as risk significant. For the most part, component level importance measures are not significantly impacted relative to the baseline SU model. Some of the extended CCF groups were impacted, and the turbine driven auxiliary feedwater interunit CCF becomes risk significant (as expected).

Unlike the two previous pilot plants, Pilot Plant C relies on an SU baseline PSA model developed using event tree linking (Pilot Plants A and B used top logic fault tree linking). Only one unit was originally modelled for the Pilot Plant C internal events, at-power PSA Level 1, because there are no major differences between the two units for Pilot Plant C. Therefore, the existing logic was duplicated and modified to generate the MUPSA model, as follows:

(1) Duplication of the existing SU model;
(2) Modification of the PSA nomenclature (e.g. BEs, gates, fault trees, event trees) to properly identify Unit 1 and Unit 2 components;
(3) Generation of a single PSA file (merging the Unit 1 and Unit 2 PSA models);
(4) Modification of the logic to account for MU interunit dependencies, such as CCF, across the site and HRA dependencies between units.

After the generation of a single PSA file by merging the Unit 1 and Unit 2 PSA models, the MUPSA model was quantified to verify that the SUCDF, number of cut sets and metric values still remained the same as the original SUPSA model. The SOS conditions can be mapped in the merged PSA model, and plant specific values were calculated for the three different SOSs based on the period 1998–2017. The general approach is illustrated in Fig. VI–4.

Because Pilot Plant C does not have cross-tied systems, no modifications in the unit dedicated systems' logic are required. Regarding shared systems, the make-up to essential service water cooling towers has the capability to supply both units simultaneously, and all of its modelled components are already

TABLE VI–8. PILOT PLANT B RESULTS FROM THE MU MODEL

SOS	SOS fraction	SU model		MU model		
		U1 CDF	U2 CDF	U1 CDF	U2 CDF	MUCDF
SOS-22	0.86	4.08E-6	4.08E-6	2.9E-06	2.9E-06	2.2E-07
SOS-12 (U1)[a]	0.07	—[b]	n.a.[c]	8.7E-07	n.a.	n.a.
SOS-12 (U2)[a]	0.07	n.a.	—[b]	n.a.	8.7E-07	
SOS-02[a]	0.00			n.a.		
Total CDF		n.a.	n.a.	3.8E-06	3.8E-06	2.2E-07

[a] LPSD model not analysed in the current investigation. Shutdown CDF not quantified.
[b] Original SU model includes (and does not distinguish between) SOS-12 and SOS-22.
[c] n.a.: not applicable.

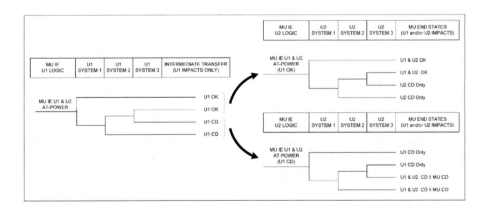

FIG. VI–4. Pilot Plant C MU event tree linking model concept.

considered common. As a consequence, the fault tree logic of this shared system is not required. In the case of the ADG, its limited capacity (only capable of supplying one emergency busbar) needs to be realistically assessed in the fault tree logic to account for prioritization and utilization of limited shared equipment. Assuming that the ADG will be required and available for Unit 1, only Unit 2 fault tree logic has been modified to account for the equipment unavailability during a SBO event. This assumption does not significantly impact results, because assuming the ADG is available for Unit 2 would just produce symmetric results from the currently obtained ones (that is, when Unit 2 is crediting the equipment from Unit 1).

A systematic review of all SU BEs was performed and new interunit CCF events were identified, as it is expected that this may impact results. Since the two units are practically identical, being functionally and physically separated and sharing very few SSCs, CCF items (especially new ones) could be the largest contributors. While potential implicit dependencies between the operator actions on different units have to be accounted for, detailed HRA modelling for MUPSAs is not included in the current phase of the investigation.

For the MUPSA model, there is a slight decrease in Unit 1 SUCDF when compared with Unit 2 SUCDF. This difference comes from the credit of the ADG only for Unit 1. Crediting the ADG represents approximately a 10% CDF decrease. As indicated before, if credit for the ADG has been given in the Unit 2 logic, the SUCDF values in the MU model would have been symmetric to those currently

TABLE VI–9. PILOT PLANT C RESULTS FROM THE MU MODEL

RISK METRIC	CDF (PER YEAR)[a]	
	SU MODEL	MU MODEL
U1 (SUCDF)	5.64E-06[b]	3.59E-07[c]
U2 (SUCDF)	5.64E-06[b]	3.94E-07[d]
U1 and U2 (MUCDF)	n.a.[f]	9.17E-09[e]

[a] Original baseline CDF = 6E-06 per year.
[b] MU-modified SUCDF quantification.
[c] Only MUIE-induced Unit 1 core damage, for SOSs when Unit 1 is operating.
[d] Only MUIE-induced Unit 2 core damage, for SOSs when Unit 2 is operating.
[e] MUIE-induced Unit 1 and Unit 2 concurrent core damage, for SOSs when both units are operating.
[f] n.a.: not applicable.

obtained (that is, Unit 1 would just have a higher SUCDF than Unit 2). Given the identical characteristics of the units, the assumption of crediting the ADG does not significantly impact the results. This is an additional SUCDF risk insight that has been captured thanks to the consideration for MU risk beyond those already assessed for SU events; this type of unavailability was not considered in the SU model. Further refinement of this sequence could be made by crediting a new local operator action accounting for dependencies between units.

A number of initial conclusions were derived from the Phase 1 effort regarding the potential technical insights and practical modelling aspects for risk-informing MU issues. Some of these conclusions are preliminary in nature and will be revisited on the basis of investigations to be performed in later phases:

(a) Existing SUPSA models may already include some design and operational aspects that are important to MU risk. In general, developing MUPSAs from existing SUPSA models requires no fundamentally new methodologies but does require expansion of existing methods and guidance to be most efficient in evaluating MU risk. Also, the process for developing a MUPSA depends on the structure of the underlying SUPSA model.

(b) Another key conclusion of this report is that the scope and extent of MUPSA modelling may be limited to specific aspects that may provide risk insights, rather than performing extensive, highly complex state of the art risk assessment. This will be used as a basis to develop a graded RIDM approach in subsequent phases of this project.

(c) In the case of cascading events, the key contribution to MUPSA will be the consideration of dependencies across multiple units (e.g. CCF, HRA), as well as potential independent events. On the latter, preliminary work indicates that, if such events are indeed independent, their overall risk contribution may be very low to justify significant modelling, to the extent that a better focus of resources may be MU initiators and those events that are driven by interunit dependencies.

(d) MU risk may be mostly dominated by specific MU initiators (e.g. MU LOOP), types of hazards (e.g. internal flooding events impacting both units, seismic events) and dependencies due to limitations in CCF and/or HRA modelling.

(e) While shared systems are important to MU risk assessment, they may also serve to substantially reduce the overall risk at some MU sites when compared with similar plants on SU sites. The beneficial effects of shared systems in SU accidents, in contrast to their contribution to MU risk, will be investigated further in later phases of this project.

(f) While new methodologies may not be necessary, the extensions of existing CCF and HRA methods are not trivial. Currently, there is limited evidence of MU CCF events in existing databases that may skew the results for CCF into conservative extrapolations for expanded CCCGs. This is expected to dominate MUPSA results for internal events and could lead to artificial risk contributors with limited risk insights (e.g. limited operational and/or maintenance changes for large numbers of components).

(g) MUPSAs may shift the risk ranking of systems and components compared with conventional PSAs, but may not drastically alter the understanding of the risk profile. Risk insights may be identified by more detailed modelling of the actions related to shared systems, cross-tied systems and shared components.

(h) While the initial investigation did not explicitly calculate the MUCDF for SOSs beyond dual-unit sites at power, the data analysis indicates that full power contributors will be the more significant MUIEs from an IE frequency perspective. This may help reduce the use of resources for estimating larger combinations of SOSs for sites with more units.

VI–3. SCOPING APPROACH FOR MUPSA

VI–3.1. Introduction

The UK ABWR site specific PSA development for the Wylfa Newydd site is being performed to capture site specific and MU aspects. Two or more reactor units are planned to be developed on the Wylfa Newydd site. It is assumed that the reactors on the site will share a common electrical grid and some shared systems and structures. Internal and external hazards have the potential to simultaneously initiate a sequence of events that could challenge the safety systems of multiple units on the site.

In support of this PSA development, GE Hitachi has developed a MUPSA methodology. This methodology was developed simultaneously to support the draft IAEA methodology report (see the body of this report). Both the UK ABWR and IAEA initiated pilot studies to demonstrate the proposed methodology. The UK ABWR results have been published [VI–30], while the IAEA pilot results are discussed in Appendix II of this publication.

MU risk results can be estimated using either a qualitative approach or a quantitative approach. There are several quantitative approaches that can be used, ranging from a scoping approach to a detailed quantitative approach. The GE Hitachi methodology developed for the ABWR PSA in the UK was a detailed approach covering all potential IEs, POSs and risk metrics. However, the level of effort estimated to develop a full Level 3 MUPSA is likely considerable without simplification and screening. An alternative to the full Level 3 PSA is to perform a simplified approach that would provide a conservative estimate for site and MU risk. The method below discusses one possible simplified approach. Although this approach focuses on core damage, the approach is applicable to Level 2 risk metrics, such as LRF.

VI–3.2. Summary of scoping approach

A scoping approach is described in Ref. [VI–31]. This approach can be applied to any risk metric, including Level 3 PSA measures. There are no specific modelling requirements, since the results are based upon SUPSA results. However, determination of the MUPSA scope is needed prior to estimating the MU risk. Additionally, the risk metrics being estimated would need to be initially identified.

This simplified approach was found to be conservative in a number of areas. In particular, the treatment of SU cascading and propagating events, which were found to be non-risk significant in the Phase II pilot activities, were treated as likely MU events in Ref. [VI–31].

A more recent approach is provided in Ref. [VI–32]. This approach screens SUIEs from the scoping assessment. Additionally, the modified scoping approach treats other MUIEs as follows:

(1) MU internal events, such as LOOP, are estimated as 10% of the total SU risk. This is applied for units without significant shared systems. For units with significant shared systems, the scoping approach assumes MU risk is equal to 100% of the SU risk.
(2) MU external events, such as seismic events, are treated as equal to the SU risk. In this case, the units are considered fully correlated.

Ref. [VI–32] defines the MU risk metrics, including MUCDF/LRF and SCDF/LRF. Details are not provided here but are summarized below. The resulting scoping approach provides the following results:

For reactor units with minimal shared systems:

$$MUCDF \leq CDF_{MUEHIE} + 0.1 \times CDF_{MUIHIE}$$
$$SCDF \leq CDF_{MUEHIE} + n \times CDF_{MUIHIE} + n \times CDF_{SUIE}$$

where n is the number of units, and MUEHIE and MUIHIE stand for a MUIE due to an external hazard or an internal hazard, respectively.

For units with significant shared systems:

$$MUCDF \leq CDF_{MUEHIE} + 1.0 \times CDF_{MUIHIE} + 0.1 \times CDF_{SUIE-MUI}$$
$$SCDF \leq CDF_{MUEHIE} + n \times CDF_{MUIHIE} + n \times CDF_{SUIE}$$

where SUIE-MUI stands for a SUIE potentially causing a MU impact.

This approach has documented limitations, including the following:

(a) The scoping approach does not consider potential risk from plants at different POSs or different sources, such as the SFP;
(b) The scoping approach does not consider unique design aspects that may impact MU risk calculation, such as plants where multiple reactors share a single containment.

If the analysis of MU risk is performed using the scoping approach and the results meet the needed goal for the site, it may be that a detailed analysis is not needed. For example, if the MUPSA is being performed to calculate a site CDF estimate and the resulting scoping analysis is below the site CDF goal, then detailed analysis is not needed. In this case, the scoping approach needs to demonstrate that the results are conservative and consider interactions among all units at the site. However, the results of the scoping approach do not support applications where the results of risk measures or similar are needed.

REFERENCES TO ANNEX VI

[VI–1]   US NUCLEAR REGULATORY COMMISSION, Title 10 Code of Federal Regulations Part 50, Appendix A, General Design Criterion 5, Sharing of Structures, Systems, and Components, US NRC, Washington, DC (1971),
https://www.nrc.gov/reading-rm/doc-collections/cfr/part050/part050-appa.html
[VI–2]   US NUCLEAR REGULATORY COMMISSION, Title 10 Code of Federal Regulations Part 100, Subsection 11, Determination of Exclusion Area, Low Population Zone, and Population Center Distance, US NRC, Washington, DC (1962),
https://www.nrc.gov/reading-rm/doc-collections/cfr/part100/part100-0011.html
[VI–3]   US NUCLEAR REGULATORY COMMISSION, US NRC Action Plan Developed as a Result of the TMI-2 Accident, NUREG-0660, US NRC, Washington, DC (1980),
https://www.nrc.gov/docs/ML0724/ML072470526.pdf
[VI–4]   US NUCLEAR REGULATORY COMMISSION, Regulatory Background on Multiunit Risk, US NRC, Washington, DC (2013),
https://www.nrc.gov/docs/ML1325/ML13255A370.pdf

[VI–5]   US NUCLEAR REGULATORY COMMISSION, Safety Goals for Nuclear Power Plant Operation, NUREG-0880, Rev. 1, US NRC, Washington, DC (1983), https://www.nrc.gov/docs/ML0717/ML071770230.pdf

[VI–6]   VIAL, E., REBOUR, V., PERRIN, B., "Severe storm resulting in partial plant flooding in 'Le Blayais' nuclear power plant", in Proc. Int. Workshop on External Flooding Hazards at Nuclear Power Plant Sites, Kalpakkam, Tamil Nadu, India, 29 Aug. – 2 Sep. 2005.

[VI–7]   US NUCLEAR REGULATORY COMMISSION, Recommendations for Enhancing Reactor Safety in the 21st Century, the Near-Term Task Force Review of Insights from the Fukushima Dai-ichi Accident, US NRC, Washington, DC (2011), https://www.nrc.gov/docs/ML1118/ML111861807.pdf

[VI–8]   INSTITUTE OF NUCLEAR POWER OPERATIONS, Lessons Learned from the Nuclear Accident at the Fukushima Daiichi Nuclear Power Station, Rev. 0, INPO 11-005 Addendum, INPO, Atlanta, GA (2012).

[VI–9]   US NUCLEAR REGULATORY COMMISSION, Memorandum: Staff Requirements — SECY-11-0124 — Recommended Actions to be Taken Without Delay from the Near-Term Task Force Report, US NRC, Washington, DC (2011), https://www.nrc.gov/docs/ML1129/ML112911571.pdf

[VI–10]   US NUCLEAR REGULATORY COMMISSION, Request for Information Pursuant to Title 10 of the Code of Federal Regulations 50.54(f) Regarding Recommendations 2.1, 2.3, and 9.3 of the Near-Term Task Force Review of Insights from the Fukushima Dai-ichi Accident, US NRC, Washington, DC (2012), https://www.nrc.gov/docs/ML1205/ML12053A340.pdf

[VI–11]   US NUCLEAR REGULATORY COMMISSION, EA-12-049, Issuance of Order to Modify Licenses with Regard to Requirements for Mitigation Strategies for Beyond-Design-Basis External Events, US NRC, Washington, DC (2012), https://www.nrc.gov/docs/ML1205/ML12054A735.pdf

[VI–12]   US NUCLEAR REGULATORY COMMISSION, Safety Goals for the Operations of Nuclear Power Plants; Policy Statement, Federal Register, Vol. 51, pp. 30028 (51 FR 30028), US NRC, Washington, DC (1986).

[VI–13]   US NUCLEAR REGULATORY COMMISSION, Feasibility Study for a Risk-Informed and Performance-Based Regulatory Structure for Future Plant Licensing, NUREG-1860, US NRC, Washington, DC (2007).

[VI–14]   US NUCLEAR REGULATORY COMMISSION, Severe Accident Risks: An Assessment of Five US Nuclear Power Plants, NUREG-1150, US NRC, Washington, DC (1990).

[VI–15]   ELECTRIC POWER RESEARCH INSTITUTE, Insights on Risk Margins at Nuclear Power Plants, Technical Evaluation of Margins in Relation to Quantitative Health Objectives and Subsidiary Risk Goals in the United States, 3002012967, EPRI, Palo Alto, CA (2018).

[VI–16]   NUCLEAR REGULATORY COMMISSION, State-of-the-Art Reactor Consequence Analyses Project Vol. 1: Peach Bottom Integrated Analysis, NUREG/CR-7110, US NRC, Washington, DC (2013), https://www.nrc.gov/reading-rm/doc-collections/nuregs/contract/cr7110/v1/

[VI–17] SCHROER, S., MODARRES, M., An event classification schema for evaluating site risk in a multi-unit nuclear power plant probabilistic risk assessment, Reliab. Eng. Syst. Saf. 117 (2013) 40–51.

[VI–18] STUTZKE, M., "Scoping estimates of multiunit accident risk", in Proc. 12th Int. Probabilistic Safety Assessment and Management Conf. (PSAM12), Honolulu, HI, 22–27 Jun. 2014,
http://meetingsandconferences.com/psam12/proceedings/paper/paper_96_1.pdf

[VI–19] NUCLEAR REGULATORY COMMISSION, Chapter 19.0 Probabilistic Risk Assessment and Severe Accident Evaluation for New Reactors, Revision 3, NUREG-0800, US NRC, Washington, DC (2015),
https://www.nrc.gov/docs/ML1508/ML15089A068.pdf

[VI–20] NUCLEAR ENERGY INSTITUTE, Modernization of Technical Requirements for Licensing of Advanced Non-Light Water Reactors: Risk-Informed Performance-Based Guidance for Non-Light Water Reactor Licensing Basis Development, NEI 18-04, Draft N, ADAMS Accession No. ML18242A469, NEI, Washington, DC (2018).

[VI–21] AMERICAN SOCIETY OF MECHANICAL ENGINEERS AND THE AMERICAN NUCLEAR SOCIETY, Probabilistic Risk Assessment Standard for Advanced Non-LWR Nuclear Power Plants, ASME/ANS RA-S-1.4–2021, Trial Use and Pilot Application, ASME/ANS, La Grange Park, IL (2021).

[VI–22] NUCLEAR REGULATORY COMMISSION, Update on Staff Plans to Apply the Full-Scope Site Level 3 PRA Project Results to the NRC's Regulatory Framework, SECY-12-0123. US NRC, Washington, DC (2012),
http://www.nrc.gov/reading-rm/doc-collections/commission/secys/2012/2012-0123scy.pdf

[VI–23] US NUCLEAR REGULATORY COMMISSION, Technical Analysis Approach Plan for Level 3 PRA Project. Rev. 0B — Working Draft, US NRC, Washington, DC (2013),
https://www.nrc.gov/docs/ML1329/ML13296A064.pdf

[VI–24] ELECTRIC POWER RESEARCH INSTITUTE, Framework for Assessing Multi-Unit Risk to Support Risk-Informed Decision-Making: Phase 1: Initial Framework Development, 3002015991, EPRI, Palo Alto, CA (2019).

[VI–25] IDAHO NATIONAL LABORATORY, Initiating Event Rates at US Nuclear Power Plants, 1988–2016, INL/EXT-17-42758, INL, Idaho Falls, ID (2017).

[VI–26] INSTITUTE OF NUCLEAR POWER OPERATIONS, INPO Consolidated Event System (ICES) Database, INPO, Atlanta, GA (accessed 2019).

[VI–27] US NUCLEAR REGULATORY COMMISSION, CCF Parameter Estimations, 2015 Update, US NRC, Washington, DC (2016),
https://nrcoe.inl.gov/resultsdb/ParamEstSpar/

[VI–28] US NUCLEAR REGULATORY COMMISSION, Guidelines on Modelling Common-Cause Failures in Probabilistic Risk Assessment, NUREG/CR-5485 (INEEL/EXT-97-01327), US NRC, Washington, DC (1998).

[VI–29] ELECTRIC POWER RESEARCH INSTITUTE, Human Reliability Analysis (HRA) for Diverse and Flexible Mitigation Strategies (FLEX) and Use of Portable Equipment: Examples and Guidance, TR-3002010663, EPRI, Palo Alto, CA (2018).

[VI–30] HIROKAWA, N., et al., "Multi-unit probabilistic safety analysis for Wylfa Newydd site", in Proc. OECD/NEA Int. Workshop on Status of Site Level PSA (including Multi-Unit PSA) Developments, Munich, 18–20 Jul. 2018.

[VI–31] STUTZKE, M., "Scoping estimates of multiunit accident risk", in Proc. 12th Int. Probabilistic Safety Assessment and Management Conf. (PSAM12), Honolulu, HI, 22–27 Jun. 2014.

[VI–32] HENNEKE, D., "Simplified methodology for multi-unit probabilistic safety assessment (PSA) modelling" (Proc. Int. Topical Mtg on Probabilistic Safety Assessment and Analysis (PSA 2019), Charleston, SC, 2019), American Nuclear Society, La Grange Park, IL (2019).

ABBREVIATIONS

AACDG	alternative AC diesel generator
ABWR	advanced boiling water reactor
AC	alternating current
ALARP	as low as reasonably practicable
ANS	American Nuclear Society
ASME	American Society of Mechanical Engineers
BE	basic event
CANDU	Canada deuterium uranium
CCCG	common cause component group
CCDP	conditional core damage probability
CCF	common cause failure
CDF	core damage frequency
COG	CANDU Owners Group
CNSC	Canadian Nuclear Safety Commission
DTA	delete-term approximation
EDG	emergency diesel generator
EPRI	Electric Power Research Institute
GDA	generic design assessment
HEP	human error probability
HFE	human failure event
HRA	human reliability analysis
ICDE	International Common Cause Failure Data Exchange
IE	initiating event
IRSN	Institut de Radioprotection et de Sûreté Nucléaire
KAERI	Korea Atomic Energy Research Institute
KHNP	Korea Hydro & Nuclear Power
LERF	large early release frequency
LMF	loss of main feedwater
LOCA	loss of coolant accident
LOOP	loss of off-site power
LPSD	low power and shutdown
LRF	large release frequency
MACCS	MELCOR Accident Consequence Code System
MCR	main control room
MCS	minimal cut set
MCUB	minimum cut upper bound
MGL	multiple Greek letter
MU	multi-unit

MUCDF	multiunit core damage frequency
MUIE	multiunit initiating event
MULRF	multiunit large release frequency
MUPSA	multiunit probabilistic safety assessment
MURCF	multiunit release category frequency
NPP	nuclear power plant
NRC	Nuclear Regulatory Commission
ONR	Office for Nuclear Regulation
PDS	plant damage state
POS	plant operational state
PRA	probabilistic risk assessment
PSA	probabilistic safety assessment
PWR	pressurized water reactor
QHO	quantitative health objectives
RC	release category
RIDM	risk informed decision making
SAM	severe accident management
SAP	Safety Assessment Principle
SBO	station blackout
SCA	safety and control area
SCDF	site core damage frequency
SFP	spent fuel pool
SFT	single fault tree
SLBO	steam line break outside containment
SOS	site operating state
SSC	structures, systems and components
STC	source term release category
SU	single unit
SUCDF	single-unit core damage frequency
SUIE	single-unit initiating event
SUPSA	single-unit PSA
TSC	technical support centre
UHS	ultimate heat sink

CONTRIBUTORS TO DRAFTING AND REVIEW

Akl, Y.	Canadian Nuclear Safety Commission (CNSC), Canada
Amico, P.	Jensen Hughes, United States of America
Apostolakis, G.	Nuclear Risk Research Centre (NRRC), Japan
Boneham, P.	Jacobsen Analytics Ltd, United Kingdom
Boulley, S.	Institut de Radioprotection et de Sûreté Nucléaire (IRSN), France
Coman, O.	International Atomic Energy Agency
Corenwinder, F.	Institut de Radioprotection et de Sûreté Nucléaire (IRSN), France
Dennis, S.	US Nuclear Regulatory Commission (US NRC), United States of America
Dermarkar, F.	CANDU Owners Group (COG), Canada
Ferrante, F.	Electric Power Research Institute (EPRI), United States of America
Fleming, K.	KNF Consulting Services LLC, United States of America
Georgescu, G.	Institut de Radioprotection et de Sûreté Nucléaire (IRSN), France
Gordon, J.	Office for Nuclear Regulation (ONR), United Kingdom
Harman, N.	Wood PLC, United Kingdom
Henneke, D.	GE Hitachi Nuclear Energy, United States of America
Hlaváč, P.	RELKO Ltd, Slovakia
Hudson, D.	US Nuclear Regulatory Commission (US NRC), United States of America

Hustak, S.	Nuclear Research Institute Řež (UJV Řež), Czech Republic
Iancu, A.	Federal Office for the Safety of Nuclear Waste Management, Germany
Jae, M.	Hanynag University, Republic of Korea
Jang, D.	Korea Institute of Nuclear Safety (KINS), Republic of Korea
Jeon, H.	Korea Hydro & Nuclear Power (KHNP), Republic of Korea
Jung, W.S.	Sejong University (Republic of Korea), Republic of Korea
Kim, D.	Korea Institute of Nuclear Safety (KINS), Republic of Korea
Kim, D.	Korea Atomic Energy Research Institute (KAERI), Republic of Korea
Kim, I.S.	NESS Ltd, Republic of Korea
Kim, M.K.	Korea Atomic Energy Research Institute (KAERI), Republic of Korea
Lee, H.	Korea Institute of Nuclear Safety (KINS), Republic of Korea
Lim, H.G.	Korea Atomic Energy Research Institute (KAERI), Republic of Korea
Maioli, A.	Westinghouse Electric Corporation, United States of America
Miura, H.	Nuclear Risk Research Centre (NRRC), Japan
Modarres, M.	University of Maryland, United States of America
Oh, K.	Korea Hydro & Nuclear Power (KHNP), Republic of Korea
Purdy, P.	Bruce Power, Canada
Poghosyan, S.	International Atomic Energy Agency

Raimond, E.	Institut de Radioprotection et de Sûreté Nucléaire (IRSN), France
Rega, J.	Tractebel, Belgium
Roewekamp, M.	Gesellschaft für Anlagen- und Reaktorsicherheit m.b.H. (GRS), Germany
Sakai, T.	Central Research Institute of Electric Power Industry (CRIEPI), Japan
Serbanescu, D.	Nuclearelectrica (SNN), Romania
Shirai, K.	Nuclear Risk Research Centre (NRRC), Japan
Vaughan, G.	Independent consultant, United Kingdom
Vayndrakh, M.	CKTI-Vibroseism, Russian Federation
Vecchiarelli, J.	Ontario Power Generation (OPG), Canada
Vida, Z.	Paks NPP, Hungary
Xu, M.	Canadian Nuclear Safety Commission (CNSC), Canada
Yaloui, S.	Canadian Nuclear Safety Commission (CNSC), Canada

Consultants Meetings

Vienna, Austria: 14–16 December 2016; 24–26 April 2017; 1–4 August 2017; 16–18 October 2017; 7–9 February 2018; 2–4 May 2018; 6–9 August 2018; 28–30 November 2018; 15–18 April 2019; 2–5 September 2019; 25–27 November 2019

Technical Meeting

Vienna, Austria: 7–10 October 2019

 **IAEA**
International Atomic Energy Agency

ORDERING LOCALLY

IAEA priced publications may be purchased from the sources listed below or from major local booksellers.

Orders for unpriced publications should be made directly to the IAEA. The contact details are given at the end of this list.

NORTH AMERICA

Bernan / Rowman & Littlefield
15250 NBN Way, Blue Ridge Summit, PA 17214, USA
Telephone: +1 800 462 6420 • Fax: +1 800 338 4550
Email: orders@rowman.com • Web site: www.rowman.com/bernan

REST OF WORLD

Please contact your preferred local supplier, or our lead distributor:

Eurospan Group
Gray's Inn House
127 Clerkenwell Road
London EC1R 5DB
United Kingdom

Trade orders and enquiries:
Telephone: +44 (0)176 760 4972 • Fax: +44 (0)176 760 1640
Email: eurospan@turpin-distribution.com

Individual orders:
www.eurospanbookstore.com/iaea

For further information:
Telephone: +44 (0)207 240 0856 • Fax: +44 (0)207 379 0609
Email: info@eurospangroup.com • Web site: www.eurospangroup.com

Orders for both priced and unpriced publications may be addressed directly to:
Marketing and Sales Unit
International Atomic Energy Agency
Vienna International Centre, PO Box 100, 1400 Vienna, Austria
Telephone: +43 1 2600 22529 or 22530 • Fax: +43 1 26007 22529
Email: sales.publications@iaea.org • Web site: www.iaea.org/publications